新工科建设之路·数据科学与大数据系列

Python实战
之数据库应用和数据获取

刘宇宙　刘　艳　◎　编著

电子工业出版社
Publishing House of Electronics Industry
北京·BEIJING

内 容 简 介

本书主要介绍 MySQL 基础、MySQL 的基本操作、MongoDB 基础和操作、文件的读写、爬虫基础、网站数据爬取实战等内容，让读者掌握通过 Python 操作关系型和非关系型数据库的方法，并通过爬虫技术获取相关数据。

本书专门针对使用 Python 操作数据库的新手或入门者，是作者学习和使用 Python 操作数据库过程的体会和经验总结。涵盖实际开发中基本的知识要点，内容详尽，代码可读性及可操作性强。

本书可以作为高等院校数据科学与大数据、计算机科学与技术等专业学生的教材，也可供财经金融管理类等专业学生参考。

未经许可，不得以任何方式复制或抄袭本书之部分或全部内容。
版权所有，侵权必究。

图书在版编目（CIP）数据

Python 实战之数据库应用和数据获取 / 刘宇宙，刘艳编著. —北京：电子工业出版社，2020.5
ISBN 978-7-121-36297-2

Ⅰ. ①P… Ⅱ. ①刘… ②刘… Ⅲ. ①软件工具－程序设计－高等学校－教材 Ⅳ. ①TP311.561

中国版本图书馆 CIP 数据核字（2019）第 068284 号

责任编辑：章海涛　　　特约编辑：穆丽丽
印　　刷：北京七彩京通数码快印有限公司
装　　订：北京七彩京通数码快印有限公司
出版发行：电子工业出版社
　　　　　北京市海淀区万寿路 173 信箱　　邮编：100036
开　　本：787×1092　1/16　印张：13.5　字数：346 千字
版　　次：2020 年 5 月第 1 版
印　　次：2024 年 8 月第 6 次印刷
定　　价：48.00 元

凡所购买电子工业出版社图书有缺损问题，请向购买书店调换。若书店售缺，请与本社发行部联系，联系及邮购电话：（010）88254888，88258888。

质量投诉请发邮件至 zlts@phei.com.cn，盗版侵权举报请发邮件至 dbqq@phei.com.cn。
本书咨询联系方式：192910558（QQ 群）。

前　言

在 2017 年和 2018 年分别出版了《Python 3.5 从零开始学》和《Python 3.7 从零开始学》两本书，很多读者朋友看后都问我是否可以再写一本 Python 进阶的书籍。起初自己并没有往这个方向继续写作的意愿，Python 进阶的书籍写起来并不容易，需要长久的技术积累才可以写出一本优质的书籍，否则容易误导读者。不过在这个思考过程中，发现并没有太多的书籍单独编写 Python 与数据库交互的内容。在慎思之下，感觉很多技术人员或许需要这样一本书，来帮助他们更好地实现 Python 与数据库的互动。特别对于那些刚入门 Python 或有一定 Python 基础的读者，这样一本专门讲解与数据库交互的图书应该对他们的技术进阶有一定的帮助。

在 2018 年，电子工业出版社找到我，询问可否出一些 Python 方面的教材，在细细思考下，列出了三本感觉可以编写的 Python 方面的书籍的提纲。

第一本是 Python 基础方面的，确定好内容提纲后，最后命名为《Python 实用教程》（ISBN 978-7-121-35884-5），主要编写 Python 基础方面的内容。该书于 2019 年由电子工业出版社出版。

第二本是本书，命名为《Python 实战之数据库应用和数据获取》，主要编写 Python 与数据库交互和数据获取的内容。

第三本书确定好内容提纲后，最后命名为《Python 实战之数据分析与处理》（ISBN 978-7-121-36347-4），将详细讲解 Numpy、Pandas 和 Matplotlib 三方面的内容，并结合这三方面的内容讲解具体的数据分析与处理的实战项目，将帮助读者向人工智能方向迈进。这本书是帮助读者通往人工智能的最后"一公里"。

这三本书将打造成一个三部曲，三本之间有一定联系。"Python 快乐学习班"将贯穿这三本书，每本书中，"Python 快乐学习班"的学生都有一个关注的主线，并以这个主线作为基础，在不同的地点完成不同知识点的学习。

《Python 实用教程》是以"Python 快乐学习班"的学生去往 Python 库游玩作为主线，在 Python 库所游玩的每个景点都与各章的知识点紧密相关，通过旅游的触景生情加深对应章节内容的理解。

本书则以数字校园作为主线，在数字校园中，"Python 快乐学习班"的学生将接触不同的数据库，学习关系型数据库（MySQL）和非关系型数据库（MongoDB）的基本知识和基本操作方式，并使用 Python 实现对不同数据库的操作。后面几章会引入数据爬虫的相关内容，最后以在音乐池中使用爬虫网爬取数据鱼作为结束。

《Python 实战之数据分析与处理》一书将以智慧城市作为主线，有兴趣的读者可以在完成本书的阅读后，前去阅读。

由于本书主要面向的是有一定 Python 基础的读者，因此本书中没有 Python 基础相关的内容。没有 Python 基础的读者可以先阅读《Python 实用教程》。当然，由其他作者编写的有关 Python 基础方面的书籍也可以。之所以推荐本人编写的图书，主要是因为本书的一些内容和本人编写的前面三本书有一定的联系，购买前面三本中的任何一本阅读，可以帮助你更快速理解本书的部分内容。

技术人员特别是编程人员掌握一定的数据库操作技术是非常有必要的。当然，研发人员不要求对数据库的操作达到大部分 DBA 的熟练程度，但需要可以完成在实现业务上对数据库的操作，就是掌握基本业务需求中的数据库操作。本书的目的是帮助没有数据库基础的读者掌握基本的数据库操作，借助一些简单项目，基本实现对数据库的增、删、改、查等操作。本书也可以帮助数据库基础薄弱的读者提升数据库的操作熟练度。

本书将关系型数据库和非关系型数据的介绍分开，先介绍关系型数据库（MySQL）的基本概念及 Python 操作 MySQL 数据库的基本示例，再介绍非关系型数据库（MongoDB）的基本概念及 Python 操作 MongoDB 的基本示例。后面具体涉及数据库操作的示例中则包含 MySQL 和 MongoDB 的操作。读者若不习惯其中一种数据库的操作方式，可以选择忽略这种数据库的操作，待有兴趣时再回看。

本书在写作过程中参考了众多参考书籍和网络资源，特别在当今的互联网环境下，很多热爱分享的技术人员在网络上分享了很多有用的信息，本书的编写过程中也从中受益颇多，在此对那些热爱分享的技术人员表示衷心的感谢。

本书的特色

本书主要介绍 MySQL 基础、MySQL 的基本操作、MongoDB 基础和操作、文件的读取、爬虫基础、网站数据爬取实战等内容，让读者掌握通过 Python 操作关系型和非关系型数据库的方法，并通过爬虫技术获取相关数据。

本书专门针对使用 Python 操作数据库的新手或刚入门者量身定做，是作者学习和使用 Python 操作数据库过程中的体会和经验总结，涵盖实际开发中基本的知识要点，内容详尽，代码可读性及可操作性强。

本书的另一个特色是，使用通俗易懂的描述和丰富的示例代码，并结合数字校园的一些事件，让本书的内容呈现尽可能生动有趣，让一些原本复杂的处理能通过另一种辅助解释得以简单化，使读者学起来轻松，充分感受到本书学习的乐趣和魅力。

本书的内容

本书共 10 章，各章内容安排如下。

第 1 章主要介绍一些数据库的发展历程及数据库的一些基本概念，为大家学习后续章节做一些铺垫。

第 2 章以 IT 大讲堂座位安排作为开端，而后主要讲解了 MySQL 的一些基本知识，及部分在实际应用中使用频率比较高的高级操作。

第 3 章以工作人员做会议安排，引出 Python 通过 PyMySQL 操作 MySQL 数据库的各种基本示例。主要讲解的是 Python 通过 PyMySQL 操作 MySQL 的各种基本示例，都是一些实战性的操作。

第 4 章网上选座示例作为开端，而后主体内容主要讲解的是 SQLAlchemy 中的一些基本概念及 SQLAlchemy 的基本操作。

第 5 章主要讲解通过 SQLAlchemy 具体操作 MySQL 数据库的一些基本使用。Python 通过 SQLAlchemy 操作 MySQL 类似学生网上选座的操作。

第 6 章主要讲解了 MongoDB 的一些基本知识，包括 MongoDB 了解、MongoDB 安装及 MongoDB 基本操作。MongoDB 的操作犹如对 mongo 山上各种植物的操作。

第 7 章主要讲解的是通过 pymongo 模块操作 MongoDB。通过 pymongo 如专业人员打理 mongo 山一样打理 MongoDB 库。

第 8 章主要讲解的是对各种文件数据的读取和写入，以及文档文件数据和关系型数据库（MySQL）及非关系型数据库（MongoDB）的交互。如选择合适的路线穿梭于数字校园一样选择合适的文档处理方式。

第 9 章主要介绍几个新的功能点及这几个新功能点的使用方式，并结合爬虫程序对这几个新功能点加以应用。简而言之，即数据爬取工具的打造。

第 10 章主要介绍通过爬取 QQ 音乐网站的数据，对前面所学内容做进一步巩固。同时通过该项目实现，了解网站数据的爬取。使用爬虫网，从音乐池中捕获数据鱼。

读者对象

有一定 Python 3.x 基础，但没有数据库基础或没有数据库实战经验的人员。
有 Python 基础，想进一步学习使用 Python 爬取数据的人员。
有一些 Python 基础后，想更进一步学习和了解 Python 3.x 应用于数据库的程序员。
Python 3.x 网课、培训机构、中学及大专院校的学生。

关于本书

由我编写清华大学出版社出版的《Python 3.5 从零开始学》《Python 3.7 从零开始学》这两本书在市场上已经获得很多读者的欢迎，《Python 实用教程》一书已由电子工业出版社出版并投入于教学。

本书的适宜人群是有一定 Python 基础，但没有数据库基础或数据库基础非常薄弱的读者，对适宜人群的学历没有任何要求，只要你有兴趣即可，当然，也希望你是可以真正看懂本书内容的读者，否则对你而言，阅读本书就是浪费时间，本书可能会给你留下一个不好的印象。

本书在编写过程中尽量以简单易懂的语言进行文本内容的组织，书中难免有编写错误的地方，希望读者可以帮助指正。写书过程中虽然自己非常小心，但是基于自身学识和技术的局限，在对一些难点的语言描述上，依然会有所偏差或错误，望读者朋友能够理解，或者告知正确的描述方式，或者给予正确的指点，也可以帮助其他读者朋友及作者少走一些弯路，本人也将致以非常诚挚的谢意。

致谢

虽然已经有了三本书的编写经验，相对于前三本书，本书编写的内容是新的知识点，所应对的也是新的人群，所以在本书的写作过程中出现了很多新的困难以及写作方式上的困惑，好在这是一个信息互联的时代，这让笔者有机会参阅很多相关信息，也让很多困难得以较好地解决。

在写作过程中参考了一些相关资源上的写作手法，这些资源上有一些技术点使用了非常形象生动的方式来阐述，参考的内容主要包括《Python 3.5 从零开始学》《Python 3.7 从零开始学》《Python 实用教程》以及 W3C 等资源。在此，对他们的编者表示真诚的感谢。

同时感谢刘艳老师，刘艳老师参与了本书几个章节的修改及校稿，对本书的一些内容做了指正和写作意见，在刘艳老师的帮助下，本书的编写进程有所提升。非常感谢刘艳老师的修改意见。

最后感谢《Python 3.5 从零开始学》《Python 3.7 从零开始学》读者们的鼓励和支持，正因为有你们通过 QQ、邮件、博客留言等方式不断要求出一本高于基础内容的书籍，才有本书的面世。也希望你们依然保持不断求知的心态，保持不断要求自己往更高方向学习的热情，往更广阔的知识海洋不断探索。

CSDN 技术博客：yuzhouliu
技术问答 E-mail：jxgzyuzhouliu@163.com
随书源码地址：http://github.com/liuyuzhou/databasesourcecode

本书相关资源

源代码

彩色原图

目 录

第 1 章 数据库简介 ·· 1
    1.1   数据库的由来 ·· 1
    1.2   数据库的发展历程 ··· 2
    1.3   数据库的作用 ·· 3
    1.4   关系型数据库 ·· 4
    1.5   非关系型数据库 ··· 6
    1.6   关系型数据库与非关系型数据库比较 ···································· 7
    1.7   小结 ··· 8

第 2 章 MySQL 的安装和操作 ··· 9
    2.1   MySQL 简介 ·· 9
    2.2   MySQL 安装 ·· 10
    2.3   MySQL 基本操作 ·· 16
        2.3.1   MySQL 基本管理 ·· 16
        2.3.2   MySQL 数据库操作 ·· 18
        2.3.3   MySQL 数据类型 ·· 19
        2.3.4   MySQL 数据表操作 ·· 21
    2.4   MySQL 高级操作 ·· 27
    2.5   小结 ··· 30
    2.6   实战演练 ··· 30

第 3 章 PyMySQL 的安装和操作 ·· 31
    3.1   PyMySQL 的介绍与安装 ··· 31
    3.2   PyMySQL 连接 MySQL 数据库 ·· 32
    3.3   PyMySQL 对 MySQL 数据库的基本操作 ···························· 34
    3.4   PyMySQL 操作多表 ·· 40
    3.5   高级封装 ··· 44
    3.6   小结 ··· 51
    3.7   实战演练 ··· 52

第 4 章 SQLAlchemy 的安装和操作 ··· 53
    4.1   SQLAlchemy 简介 ·· 53
    4.2   SQLALchemy 的安装和连接 ·· 54
        4.2.1   安装 SQLAlchemy ·· 54
        4.2.2   使用 SQLAlchemy 连接 MySQL 数据库 ·················· 55

		4.2.3 映射声明	56
4.3	SQLAlchemy 常用数据类型		56
4.4	创建类		57
4.5	创建模式		59
4.6	创建映射类的实例		61
4.7	创建会话		62
4.8	小结		64
4.9	实战演练		64

第 5 章 SQLAlchemy 操作 MySQL ... 65

5.1	SQLAlchemy 对 MySQL 数据库的基本操作		65
		5.1.1 添加对象	65
		5.1.2 查询对象	70
		5.1.3 更新对象	72
		5.1.4 删除对象	74
5.2	SQLAlchemy 的常用 filter 操作符		75
		5.2.1 equals 操作符	76
		5.2.2 not equals 操作符	76
		5.2.3 like 操作符	77
		5.2.4 and 操作符	78
		5.2.5 or 操作符	78
		5.2.6 is null 操作符	78
		5.2.7 is not null 操作符	78
		5.2.8 in 操作符	79
		5.2.9 not in 操作符	79
5.3	SQLAlchemy 的更多操作		79
		5.3.1 返回列表和单项	79
		5.3.2 嵌入使用 SQL	80
		5.3.3 计数	82
5.4	小结		83
5.5	实战演练		84

第 6 章 MongoDB 基础 ... 85

6.1	MongoDB 简介		85
6.2	MongoDB 的安装		86
6.3	MongoDB 基本概念		92
		6.3.1 文档	92
		6.3.2 集合	93
		6.3.3 数据库	93
		6.3.4 数据类型	94

6.4 MongoDB 基本操作 ··· 95
    6.4.1 创建数据库 ·· 95
    6.4.2 删除数据库 ·· 96
    6.4.3 创建集合 ··· 97
    6.4.4 删除集合 ··· 98
    6.4.5 插入文档 ··· 98
    6.4.6 更新文档 ··· 99
    6.4.7 删除文档 ·· 101
    6.4.8 MongoDB 查询文档 ·· 103
    6.4.9 条件操作符 ·· 105
    6.4.10 $type 操作符 ·· 107
    6.4.11 limit()和 skip()方法 ·· 108
    6.4.12 排序 ··· 109
    6.4.13 索引 ··· 109
    6.4.14 聚合 ··· 111
6.5 小结 ·· 112
6.6 实战演练 ··· 112

第 7 章 Python 操作 MongoDB ··· 113
7.1 pymongo 安装 ·· 113
7.2 Python 连接 MongoDB ·· 113
7.3 Python 对 MongoDB 的基本操作 ··· 114
    7.3.1 创建数据库 ·· 115
    7.3.2 创建集合 ·· 115
    7.3.3 查询文档 ·· 116
    7.3.4 插入文档 ·· 119
    7.3.5 更改文档 ·· 122
    7.3.6 文档排序 ·· 125
    7.3.7 删除文档 ·· 127
7.4 小结 ·· 131
7.5 实战演练 ··· 131

第 8 章 文件读写 ·· 132
8.1 with 语句 ··· 132
8.2 TXT 文件读写 ·· 133
8.3 CSV 文件读写 ·· 136
8.4 JSON 文件读写 ··· 139
8.5 Word 文件读写 ··· 141
8.6 XML 文件读取 ·· 144
8.7 CSV 文件读取后插入 MySQL 数据库 ·· 145

8.8	CSV 文件读取后插入 MongoDB 数据库	150
8.9	小结	154
8.10	实战演练	154

第 9 章 Python 数据爬取 ... 155

9.1	爬虫基础	155
9.2	库的安装与使用	156
	9.2.1 pyecharts 库的安装与使用	156
	9.2.2 jieba 分词库的安装与使用	157
	9.2.3 BeautifulSoup 库的安装与使用	157
	9.2.4 Requests 库的安装与使用	158
9.3	分词与词频统计实战	158
	9.3.1 整体结构设计	159
	9.3.2 数据结构设计	159
	9.3.3 数据的爬取与保存	160
	9.3.4 制定关键词库	162
	9.3.5 词频统计与图表生成	162
9.4	分词和词频统计的完整代码	164
9.5	小结	172
9.6	实战演练	172

第 10 章 项目实战：音乐数据爬取 ... 173

10.1	获取全部歌手	173
10.2	获取歌手的歌曲数目	178
10.3	获取每首歌曲信息	184
10.4	歌曲下载	186
10.5	歌曲信息持久化	192
10.6	可视化展示	198
10.7	小结	200

附录 A MySQL 的四个默认库 ... 201

附录 B PyMySQL 连接对象全量参数解释 ... 205

第 1 章　数据库简介

即使没有接触过数据库，读者对"数据库"这个词应该并不陌生。本章将简单介绍数据库的一些基本概念，包括数据库的由来、关系型数据库和非关系型数据库等。

1.1　数据库的由来

在开始介绍数据库前，先请读者了解数据库的由来。

数据库的历史可以追溯到 50 多年前，那时的数据管理非常简单。其操作方式为通过机器运行数百万张穿孔卡片来进行数据处理，其运行结果在纸上打印出来或者制成新的穿孔卡片。而数据管理就是对所有这些穿孔卡片进行物理的存储和处理。

数据库的产生背景是美国为了在战争中保存情报资料。就如计算机的出现，是因为战争而演化出来的一样，阿波罗登月计划对数据库的发展起到了推动作用。

在数据库进入民用后，科学家在理论上进行研究，并发表了论文，对数据库的发展起到了理论支持的作用。

首先使用"DataBase"一词的是美国系统发展公司。该公司在 20 世纪 60 年代为美国海军基地研制数据时，研发人员在技术备忘录中首先使用了 DataBase 一词。

数据库的发展经历了人工管理阶段、文件系统阶段、数据库阶段和高级数据库阶段。

1. 人工管理阶段

刚开始没有硬盘等存储器，用的是纸带等进行数据的存储。编程人员在写程序的时候，通常需要根据数据来编写程序，同时需要考虑到数据的物理存储结构。在那个时候对程序员的要求是非常高的，程序员的负担也非常重，但是效率非常低。那时还没有"文件"的概念，"文件"的概念被引入后对数据库的推动作用非常巨大。

2. 文件系统阶段

这个阶段引入了"文件"的概念，数据存储在文件中，逻辑结构和物理结构有所区分，虽然是一种不够彻底的结构，但有如下明显的优点：文件的组织具有多样化特点；数据可以重复使用；对数据操作的颗粒比较大，以记录为单位。

这是人类史上一个巨大的进步，因为从此开始，人类有了"存储"概念，有了分离物理和逻辑的思维，这也是后续关系型数据库出现的一个重大铺垫。为什么呢？因为文件很容易导致数据的冗余，而冗余进一步导致了数据的不一致，还有一个问题就是数据间的联系弱。为了解决这些，科研人员研发出了数据模型。在提出数据模型后，文件系统的那些问题就解决了。

3. 数据库阶段

数据模型和数据结构的发展可以解决文件系统的问题。为了使数据更加实用，继而发展了数据控制技术，并随着数据控制技术不断成熟，数据库在实际生活中也逐步被广泛应用起来。

4. 高级数据库阶段

在数据库的实际应用中，计算机技术和网络技术中不断产生新问题和新挑战，在这些新需求的刺激下，产生了分布式数据库、面向对象数据库和网络数据库等更高级的数据库。分布式数据库可以解决集中管理带来的过度复杂、拥挤的数据处理问题。面向对象数据库解决了多媒体数据、多维表格数据、CAD 数据的表达等问题。

1.2 数据库的发展历程

数据库的发展大概经历了摇篮和萌芽阶段、发展阶段和成熟阶段，各个阶段的主要事迹及发生年份对应如下。

1. 摇篮和萌芽阶段

数据库系统的萌芽出现于 20 世纪 60 年代。当时计算机开始广泛地应用于数据管理，对数据的共享提出了越来越高的要求。

由于处理的需求越来越多，传统的文件系统已经不能满足人们的需要。能够统一管理和共享数据的数据库管理系统（DBMS）应运而生。

1963 年，C.W. Bachman 设计开发的 IDS（Integrate Data Store）系统开始投入运行，它可以为多个 COBOL 程序共享数据库。这是世界上第一个网状数据库管理系统，也是第一个数据库管理系统——集成数据存储（Integrated Data Store，IDS）。IDS 奠定了网状数据库的基础，并在当时得到了广泛的发行和应用。

1968 年，网状数据库系统 TOTAL 等开始出现。网状数据库模型对于层次和非层次结构的事物都能比较自然地模拟，在关系型数据库出现之前，网状数据库比层次数据库用得普遍。在数据库发展史上，网状数据库占有重要地位。

1969 年，IBM 的 Mc Gee 等人开发的层次数据库系统（Information Management System，IMS）发布，可以让多个程序共享数据库。IMS 是最著名最典型的层次数据库系统，是一种适合其主机的层次数据库，是 IBM 公司研制的最早的大型数据库系统程序产品。

1969 年 10 月，CODASYL 数据库研制者提出了网状数据库系统规范报告 DBTG，使数据库系统开始走向规范化和标准化，所以许多专家认为数据库技术起源于 20 世纪 60 年代末。

数据库技术的产生来源于社会的实际需要，而数据技术的实现必须有理论作为指导，系统的开发和应用又不断地促进数据库理论的发展和完善。

2. 发展阶段

20 世纪 80 年代，大量商品化的关系型数据库系统问世并被广泛地推广使用。这时的数据库系统既有适用于大型计算机系统的，也有适用于中、小型和微型计算机系统的。大概在这个时期，分布式数据库系统也逐步使用。

1970 年，IBM 公司 San Jose 研究所的 E.F. Code 发表了题为《大型共享数据库的数据关系模型》的论文，由此开创了数据库的关系方法和关系规范化的理论研究。

关系方法由于其理论上的完美和结构上的简单，对数据库技术的发展起了至关重要的作用，成功地奠定了关系数据理论的基石。

1971 年，美国数据系统语言协会在正式发表的 DBTG 报告中提出了三级抽象模式，解决了数据独立性的问题，即：对应用程序所需的数据结构描述的外模式，对整个客体系统数据结构描述的概念模式，对数据存储结构描述的内模式。

1974 年，IBM 公司 San Jose 研究所成功研制了关系型数据库管理系统 System R，并且投放到软件市场。

1976 年，美籍华人陈平山提出了数据库逻辑设计的实际（体）联系方法。

1978 年，新奥尔良发表了 DBDWD 报告，他把数据库系统的设计过程划分为四个阶段：需求分析、信息分析与定义、逻辑设计和物理设计。

1980 年，J.D. Ulman 所著《数据库系统原理》一书出版。

1981 年，E.F. Code 获得了计算机科学的最高奖——ACM 图灵奖。

1984 年，David Marer 所著《关系型数据库理论》一书出版，标志着数据库在理论上的成熟。

3．成熟阶段

20 世纪 80 年代至今，数据库理论和应用进入成熟发展时期。关系型数据库和非关系型数据库都得到快速发展和广泛应用。特别伴随着互联网、云计算技术的发展，各项数据库技术也得到了迅猛的发展。

1.3 数据库的作用

为什么需要了解数据库，数据库的作用是什么？

1．实现数据共享

数据共享包括所有用户可同时存取数据库中的数据，也包括用户可以用各种方式通过接口使用数据库，并提供数据共享。

2．减少数据的冗余度

同文件系统相比，由于数据库实现了数据共享，从而避免了用户各自建立应用文件。数据库减少了大量重复数据，减少了数据冗余，维护了数据的一致性。

3．实现数据的独立性

数据的独立性包括逻辑独立性（数据库中数据的逻辑结构与应用程序相互独立）和物理独立性（数据物理结构的变化不影响数据的逻辑结构）。

4．实现数据集中控制

文件管理方式中，数据处于一种分散的状态，不同的用户或同一用户在不同处理中其文件之间毫无关系。利用数据库可对数据进行集中控制和管理，并通过数据模型表示各种数据的组织以及数据间的联系。

5．数据一致性和可维护性，确保数据的安全性和可靠性

① 安全性控制：防止数据丢失、错误更新和越权使用。

② 完整性控制：保证数据的正确性、有效性和相容性。

③ 并发控制：在同一时间周期内，允许对数据实现多路存取，又能防止用户之间的不正常交互作用。

6. 故障恢复

由数据库管理系统提供一套方法，可及时发现故障和修复故障，从而防止数据被破坏。数据库系统能尽快恢复数据库系统运行时出现的故障，可能是物理上或是逻辑上的错误，如对系统的误操作造成的数据错误等。

数据库有很多类型，从最简单的存储有各种数据的表格到能够进行海量数据存储的大型数据库系统，在各方面得到了广泛的应用。

在信息化社会，充分、有效地管理和利用各类信息资源，是进行科学研究和决策管理的前提条件。数据库技术是管理信息系统、办公自动化系统、决策支持系统等信息系统的核心部分，是进行科学研究和决策管理的重要技术手段。

数据库技术发展到现在，已经非常成熟了。当前主流的有关系型数据库和非关系型数据库两种，1.4 节和 1.5 节会分别进行介绍。

1.4 关系型数据库

关系型数据库是建立在关系模型基础上的数据库，借助集合代数等数学概念和方法来处理数据库中的数据。现实世界中的各种实体以及实体之间的各种联系均用关系模型来表示。

关系型数据库以行和列的形式存储数据，以便于用户理解。这一系列的行和列被称为表，一组表组成了数据库。

关系型数据库并不是唯一的高级数据库模型，也不是性能最优的模型，但是关系模型是现今使用最广泛、最容易理解和使用的数据库模型。所谓关系型数据库，是指采用了关系模型来组织数据的数据库。

简单来说，关系模型指的是二维表格模型，关系型数据库就是由二维表及其之间的联系组成的一个数据组织。

关系模块中常用的操作包括：选择、投影、连接、并、交、差、除、数据操作、增加、删除、修改、查询等。

关系型数据库的完整性包括：实体完整性、参照完整性、用户定义完整性。

关系模型中的常用概念如下。

① 关系：可以理解为一张二维表，每个关系都具有一个关系名，就是通常说的表名。

② 元组：可以理解为二维表中的一行，在数据库中经常被称为记录。

③ 属性：可以理解为二维表中的一列，在数据库中经常被称为字段。

④ 域：属性的取值范围，即数据库中某一列的取值限制。

⑤ 关键字：一组可以唯一标识元组的属性，在数据库中常被称为主键，由一个或多个列组成。

⑥ 关系模式：指对关系的描述，其格式为：关系名(属性 1, 属性 2, …, 属性 N)。在数据库中通常称为表结构。

1. 关系型数据库的优点

① 容易理解：二维表结构是非常贴近逻辑世界的一个概念，关系模型相对网状、层次等其他模型来说更容易理解。

② 使用方便：通用的 SQL 使得操作关系型数据库非常方便，程序员甚至于数据库管理员可以方便地在逻辑层面操作数据库，而完全不必理解其底层实现。

③ 易于维护：丰富的完整性（实体完整性、参照完整性和用户定义的完整性）大大降低了数据冗余和数据不一致的概率。

2. 关系型数据库的准则

关系型数据库系统应该完全支持关系模型的所有特征。关系模型的奠基人埃德加·科德具体地给出了关系型数据库应遵循的基本准则。具体如下：

准则 0：关系型数据库管理系统必须能完全通过它的关系能力来管理数据库。

准则 1：信息准则。关系型数据库管理系统的所有信息都应该在逻辑一级上用表中的值这种方法显式地表示。

准则 2：保证访问准则。依靠表名、主键和列名的组合，保证能以逻辑方式访问关系型数据库中的每个数据项。

准则 3：空值的系统化处理。全关系的关系型数据库管理系统支持空值的概念，并用系统化的方法处理空值。

准则 4：基于关系模型的动态的联机数据字典。数据库的描述在逻辑级上和普通数据采用同样的表述方式。

准则 5：统一的数据子语言。关系型数据库管理系统可以具有几种语言和多种终端访问方式，但必须有一种语言，它的语句可以表示为严格语法规定的字符串，并能全面地支持各种规则。

准则 6：视图更新准则。所有理论上可更新的视图也应该允许由系统更新。

准则 7：高级的插入、修改和删除操作。系统应该对各种操作进行查询优化。

准则 8：数据的物理独立性。无论数据库的数据在存储表示或访问方法上做任何变化，应用程序和终端活动都保持逻辑上的不变性。

准则 9：数据逻辑独立性。当对基本关系进行理论上信息不受损害的任何改变时，应用程序和终端活动都保持逻辑上的不变性。

准则 10：数据完整的独立性。关系型数据库的完整性约束条件必须是用数据库语言定义并存储在数据字典中的。

准则 11：分布独立性。关系型数据库管理系统在引入分布数据或数据重新分布时保持逻辑不变。

准则 12：无破坏准则。如果关系型数据库管理系统使用低级语言，那么这个低级语言不能违背或绕过完整性准则。

当前主流的关系型数据库有 Oracle、DB2、Microsoft SQL Server、Microsoft Access、MySQL 等。

1.5 非关系型数据库

非关系数据库通常被称为 NoSQL 数据库。NoSQL（NoSQL=Not Only SQL），意即"不仅仅是 SQL"，是一项全新的数据库革命性运动，早期就有人提出，2009 年开始被高度关注并广泛使用。

随着互联网 Web 2.0 网站的兴起，传统的关系型数据库在应付 Web 2.0 网站，特别是超大规模和高并发的 SNS 类型的 Web 2.0 纯动态网站时，已经显得力不从心，暴露了很多难以克服的问题，所以非关系型数据库由于其本身的特点得到了非常迅速的发展。

NoSQL 数据库的产生就是为了解决大规模数据集合多重数据种类带来的挑战，尤其是在大数据应用方面。

虽然 NoSQL 流行起来才几十年的时间，但是不可否认，现在已经开始了第二代运动。尽管早期的堆栈代码只能算是一种试验，然而现在的系统已经更加成熟、稳定。不过现在也面临着一个严酷的事实：技术越来越成熟——以致原来很好的 NoSQL 数据存储不得不进行重写，也有少数人认为这就是所谓的 2.0 版本。该工具可以为大数据建立快速、可扩展的存储库。

NoSQL 数据库的分类如下。

1．键值（key-value）存储数据库

键值存储数据库主要使用哈希表，其中有一个特定的键和一个指针指向特定的数据。key-value 模型对于 IT 系统的优势在于简单、易部署。如果 DBA 只对部分值进行查询或更新，key-value 模型就显得效率低下了。

这类数据库包括：Tokyo Cabinet/Tyrant、Redis、Voldemort、Oracle BDB。

2．列存储数据库

列存储数据库通常用来应对分布式存储的海量数据。键仍然存在，但是它们的特点是指向了多列。这些列是由列家族来安排的。

列存储数据库包括 Cassandra、HBase、Riak。

3．文档型数据库

文档型数据库的灵感是来自 Lotus Notes 办公软件，同键值存储数据库类似。其数据模型是版本化的文档，半结构化的文档以特定的格式存储，如 JSON。文档型数据库可以看作是键值数据库的升级版，允许嵌套键值，但比键值存储数据库的查询效率更高。

文档型数据库包括：CouchDB、MongoDB、SequoiaDB（已经开源）。

4．图形（Graph）数据库

图形数据库同其他行列及刚性结构的 SQL 数据库不同，使用灵活的图形模型，并且能够扩展到多个服务器上。NoSQL 数据库没有标准的查询语言（SQL），因此进行数据库查询需要创建数据模型。许多 NoSQL 数据库都有 REST 式的数据接口或者查询 API。

图形数据库包括：Neo4J，InfoGrid，Infinite Graph。

NoSQL 数据库在以下几种情况下比较适用：

① 数据模型比较简单。
② 需要灵活性更强的 IT 系统。

③ 对数据库性能要求较高。
④ 不需要高度的数据一致性。
⑤ 对于给定 key，比较容易映射复杂值的环境。

1.6 关系型数据库与非关系型数据库比较

在 1.4 节和 1.5 节大概讲解了关系型数据库和非关系型数据的概念和一些应用场景。学习后你可能会有疑问：关系型数据库和非关系型数据库使用哪一个好，该怎么选择？

首先来了解关系型数据库的理论——ACID。ACID 是指数据库管理系统（DBMS）在写入或更新资料的过程中，为保证事务（transaction）是正确可靠的，所必须具备的四个特性：原子性（Atomicity，或称不可分割性）、一致性（Consistency）、隔离性（Isolation，又称独立性）、持久性（Durability）。

1．A（Atomicity）：原子性

一个事务（transaction）中的所有操作，要么全部完成，要么全部不完成，不会结束在中间某个环节。事务在执行过程中发生错误，会被回滚（Rollback）到事务开始前的状态，就像这个事务从来没有被执行过一样。

2．C（Consistency）：一致性

在事务开始之前和事务结束以后，数据库的完整性没有被破坏。这表示写入的资料必须完全符合所有的预设规则，包括资料的精确度、串联性以及后续数据库可以自发地完成预定的工作。

3．I（Isolation）：隔离性

数据库允许多个并发事务同时对其数据进行读写和修改，隔离性可以防止多个事务并发执行时由于交叉执行而导致数据的不一致。事务隔离分为不同级别，包括读未提交（Read Uncommitted）、读提交（Read Committed）、可重复读（Repeatable Read）和串行化（Serializable）。

4．D（Durability）：持久性

事务处理结束后，对数据的修改就是永久的，即便系统故障也不会丢失。

关系型数据库严格遵循 ACID 理论。但当数据库要开始满足横向扩展、高可用、模式自由等需求时，需要对 ACID 理论进行取舍，不能严格遵循 ACID。

关系型数据库的优势如下：
① 保持数据的一致性（事务处理）。
② 由于以标准化为前提，数据更新的开销很小（相同的字段基本上只有一处）。
③ 可以进行 JOIN 等复杂查询。

其中能够保持数据的一致性是关系型数据库的最大优势。

关系型数据库的不足如下：
① 不擅长大量数据的写入处理。
② 不能为有数据更新的表做索引或表结构（schema）变更。
③ 字段不固定时可用性比较差。

④ 不擅长对简单查询需要快速返回结果的处理。

大量数据的写入处理会使读写集中在一个数据库上，让数据库不堪重负。大部分网站已使用主从复制技术实现读写分离，以提高读写性能和读库的可扩展性。

在进行大量数据操作时，会使用数据库主从模式。数据的写入由主数据库负责，数据的读入由从数据库负责，可以比较简单地通过增加从数据库来实现规模化，但是数据的写入却没有简单的方法来解决规模化问题。

非关系型数据库（NoSQL）的优势如下：

① 成本低。NoSQL 数据库简单易部署，基本都是开源软件，不需像使用 Oracle 那样花费大量成本购买，比关系型数据库价格便宜。

② 查询速度快。NoSQL 数据库将数据存储于缓存中，关系型数据库将数据存储在硬盘中，查询速度远不及 NoSQL 数据库。

③ 存储数据的格式灵活。NoSQL 的存储格式是 key-value 形式、文档形式、图片形式等，所以可以存储基础类型、对象、集合等格式，而数据库只支持基础类型。

④ 易扩展。NoSQL 数据库种类繁多，其共同点是去掉了关系型数据库的关系特性。数据之间无关系，这样非常容易扩展，同时在架构的层面上带来了可扩展的能力。

⑤ 灵活的数据模型。NoSQL 不需事先为要存储的数据建立字段，随时可以存储自定义的数据格式。而在关系型数据库中，增、删字段是一件非常麻烦的事情。如果是非常大数据量的表，增加字段简直是噩梦。这在大数据量的 Web 2.0 时代尤其明显。

⑥ 高可用。NoSQL 在不太影响性能的情况下，可以方便地实现高可用的架构。比如，Cassandra、HBase 模型通过复制也能实现高可用。

非关系型数据库（NoSQL）的缺点如下：

① 维护的工具和资料有限。因为 NoSQL 是属于新的技术，不能与关系型数据库几十年的技术积累同日而语。

② 不提供对 SQL 的支持。如果不支持 SQL 这样的工业标准，将产生一定用户的学习和使用成本。

③ 不提供关系型数据库那样对事务的处理。

④ 没有存储过程。NoSQL 数据库中大多没有存储过程。

⑤ 支持的特性不够丰富，产品不够成熟。现有产品所提供的功能都比较有限，不像 Microsoft SQL Server 和 Oracle 那样能提供各种附加功能，如 BI 和报表等。大多数产品还处于初创期，与关系型数据库几十年的完善不可同日而语。

当然，随着 NoSQL 的应用越来越广泛，投入 NoSQL 研发的人员越来越多，NoSQL 技术在最近几年也得到了迅猛发展，很多应用上的不足正在逐步被完善。

1.7 小结

本章主要介绍数据库的发展历程及数据库的基本概念，为大家学习后续章节做一些铺垫。

本书将以 MySQL 作为关系型数据库的代表对象展开讲解，以 MongoDB 作为非关系型数据库的代表对象展开讲解。第 2 章将讲解 MySQL 的基本知识及操作。

第 2 章 MySQL 的安装和操作

MySQL 作为一个开源数据库，是关系型数据库的典型代表。本章先从 MySQL 的基本概念开始，然后介绍 MySQL 的安装和操作。

数字校园为提高学生们对人工智能方面的认识，请来某专家为同学们讲解相关知识，讲座地点在 IT 大讲堂。IT 大讲堂的座位都是固定的，这样便于座位的分配。由于会场可容纳人数有限，每个班只有 10 个名额。"Python 快乐学习班"的名额也只有 10 个，他们班分配的座位号是 A 区 I 排 1～10 号的位置。对于去 IT 大讲堂听讲座的同学，不用担心是否有位置，位置已经都安排好了，大家对号入座即可，不用抢位，也不用占位。如对于 A 区 I 排 1 到 10 号的位置，"Python 快乐学习班"的同学只管大胆坐上即可，他们若坐到其他位置，那就要随时做好被人赶走的准备。其他同学也不用担心是否有位置，因为他们根本没有被安排位置。

关系型数据库的操作与分配座位类似，一旦定义了表结构，对插入表中的数据格式就确定了，都需要保持一致，不一致的就没有资格进入数据库。

2.1 MySQL 简介

MySQL 是目前最流行、使用最多的关系型数据库管理系统（Relational Database Management System，RDBMS）。在 Web 应用方面，MySQL 是最好的 RDBMS 应用软件之一。

MySQL 由瑞典 MySQL AB 公司开发，目前属于 Oracle 公司。MySQL 将数据保存在不同的表中，而不是将所有数据放在一个大仓库中，这样增加了数据处理速度并提高了灵活性。

MySQL 数据库的特性如下：

① MySQL 是开源的。MySQL 并不需要支付额外费用。
② MySQL 支持大型的数据库，可以处理拥有上千万条记录的大型数据库。
③ MySQL 使用标准的 SQL 形式，与很多收费数据库软件有相同的标准。
④ MySQL 可以运行于多个系统上，并支持多种语言。这些语言包括 C、C++、Python、Java、Perl、PHP 和 Ruby 等。
⑤ 32 位 MySQL 系统表文件最大可支持 4 GB，64 位 MySQL 系统支持最大的表文件为 8 TB。
⑥ MySQL 可以定制，采用 GPL 协议。用户可以修改源码开发自己的 MySQL。

关系型数据库管理系统的一些术语如下。

数据库：一些关联表的集合，可理解为存放关联表的仓库。
表：数据的矩阵。在一个数据库中，表看起来像一个简单的电子表格。
列：一列（数据元素）包含了相同的数据，如邮政编码的数据。

行：一行（即元组或记录）是一组相关的数据，如一条用户订阅的数据。

冗余：存储两倍或以上数据，冗余降低了性能，但提高了数据的安全性。

主键：指一列或多列的组合，其值能唯一地标识表中的每一行，从而强制表的实体完整性。主键主要用于与其他表的外键关联，以及本记录的修改和删除。

外键：如果公共关键字在一个关系中是主关键字，那么这个公共关键字被称为另一个关系的外键。外键用于关联两个表。

复合键（组合键）：将多列作为一个索引键，一般用于复合索引。

索引：用于快速访问数据库表中的特定信息。索引是对数据库表中一列或多列的值进行排序的一种结构，类似书籍的目录。

参照完整性：要求关系中不允许引用不存在的实体。与实体完整性是关系模型必须满足的完整性约束条件，目的是保证数据的一致性。

表头（header）：每一列的名称。

值（value）：行的具体信息，每个值必须与该列的数据类型相同。

键（key）：键的值在当前列中具有唯一性。

MySQL 为关系型数据库，这种"关系型"可以理解为"表格"的概念，一个关系型数据库由一个或数个表格组成。

2.2 MySQL 安装

本书以 MySQL 8.0.11 版本作为示例进行讲解。MySQL 的官方网址为：https://www.mysql.com/，官网首页如图 2-1 所示。

图 2-1 MySQL 官网首页

单击图 2-1 中的"DOWNLOADS"，即可进入下载页面：https://dev.mysql.com/downloads/mysql/，如图 2-2 所示。单击其中的下拉框，可以选择适合各操作系统的 MySQL，如图 2-3 所示。如果选择 Microsoft Windows 操作系统（如图 2-4 所示），则出现相应软件的压缩包，然后单击"Download"按钮即可开始下载。如果选择 macOS 操作系统，则出现如图 2-5 所示的页面，然后单击"Download"按钮，出现如图 2-6 所示的页面，直接单击底部地"No thanks, just start my download."，即可开始软件下载。

图 2-2 MySQL 下载（一）

图 2-3 MySQL 下载（二）

图 2-4 MySQL 下载（三）

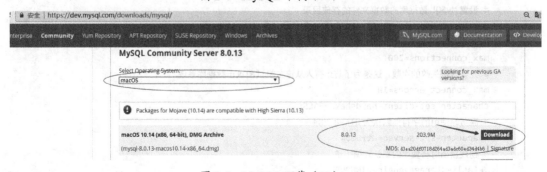

图 2-5 MySQL 下载（四）

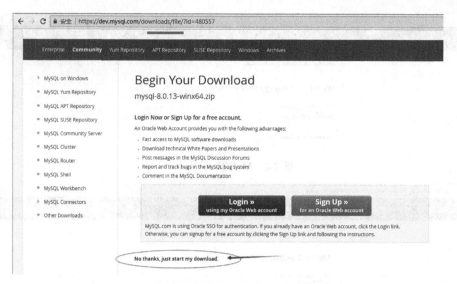

图 2-6 MySQL 下载（五）

下面介绍 Windows 下 MySQL 8.0.11 版本的安装。

1．解压 ZIP 包到指定目录

MySQL 8.0.11 的安装包为 ZIP 文件，解压到指定目录，如示例中的安装目录是 E:\mysql\mysqlinstall。解压后得到的全路径为 E:\mysql\mysqlinstall\mysql-8.0.11-winx64。解压后的文件夹情况如图 2-7 所示。

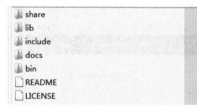

图 2-7 解压结果

2．添加配置文件

解压后的目录中并没有 my.ini 文件，需要自行创建。在安装根目录下添加 my.ini，示例路径为 E:\mysql\mysqlinstall\mysql-8.0.11-winx64\my.ini。写入基本配置：

```
[mysqld]
# 设置 3306 端口
port=3306
# 设置 MySQL 的安装目录
basedir=E:/mysql/mysqlinstall/mysql-8.0.11-winx64
# 设置 MySQL 数据库的数据文件的存放目录
datadir=E:/mysql/mysqlinstall/mysql-8.0.11-winx64/Data
# 允许最大连接数
max_connections=200
# 允许连接失败的次数。这是为了防止有人从该主机试图攻击数据库系统
max_connect_errors=10
character-set-client-handshake=FALSE
# 服务端使用的字符集默认为 UTF-8 MB4
character-set-server=utf8mb4
# 创建新表时将使用的默认存储引擎
default-storage-engine=INNODB
collation_server=utf8mb4_unicode_ci
[mysql]
```

```
# 设置 MySQL 客户端默认字符集
default-character-set=utf8mb4
[client]
# 设置 MySQL 客户端连接服务端时默认使用的端口
port=3306
default-character-set=utf8mb4
```

注意：basedir 是示例中本地的安装目录，datadir 是示例中数据库数据文件要存放的位置，各项配置需要根据自己的环境进行配置。

查看所有配置项的具体用途及含义，可参考如下网址：

https://dev.mysql.com/doc/refman/8.0/en/mysqld-option-tables.html

3. 数据库初始化

配置好 my.ini 文件后，接下来做数据库的初始化。

进入 MySQL 安装目录的 bin 目录下执行以下命令：

```
mysqld --initialize –console
```

执行以上命令后，会打印 root 用户的初始密码，完整操作如下：

```
C:\Users\lyz>cd D:\install\mysql-8.0.11-winx64\bin
E:\mysql\mysqlinstall\mysql-8.0.11-winx64\bin>mysqld --initialize --console
2018-11-12T12:12:51.573693Z 0 [System] [MY-013169] [Server] E:\mysql\mysqlinstall\
mysql-8.0.11-winx64\bin\mysqld.exe (mysqld 8.0.11) initializing of server in progress as
process 13604
2018-11-12T12:13:13.988826Z 5 [Note] [MY-010454] [Server] A temporary password is generated
for root@localhost: h*Fc#8FhuufA
2018-11-12T12:13:26.366886Z 0 [System] [MY-013170] [Server] E:\mysql\mysqlinstall\
mysql-8.0.11-winx64\bin\mysqld.exe (mysqld 8.0.11) initializing of server has completed

E:\mysql\mysqlinstall\mysql-8.0.11-winx64\bin>
```

注意，执行输出结果里面有一段：[Note] [MY-010454] [Server] A temporary password is generated for root@localhost: h*Fc#8FhuufA。在这个输出结果中，root@localhost:后面的"h*Fc#8FhuufA"是初始密码（不含首位空格），在没有更改密码前，需要记住这个密码，后续操作中的登录需要用到。

如果执行到这一步时忘记了密码可以到安装目录下，将 datadir 配置的目录删除，再执行一遍初始化命令，就可以重新生成对应的密码。

4. 安装服务

执行完上面的操作命令后，就可以通过命令 net start mysql 启动 MySQL 的服务了。

具体操作示例如下：

```
E:\mysql\mysqlinstall\mysql-8.0.11-winx64\bin>net start mysql
MySQL 服务正在启动 ..
MySQL 服务已经启动成功。
```

当看到上面的输出命令时，即代表 MySQL 已经安装成功，此时 MySQL 服务也已经启动。

5. 更改密码

MySQL 服务启动后，就可以进入 MySQL 的命令模式了，进入的指令形式如下：

```
mysql -h 主机名 -u 用户名 -p
```

参数说明如下。

-h：指定客户端所要登录的 MySQL 主机名，登录本机（localhost 或 127.0.0.1）时，本参数可以省略。

-u：登录的用户名。

-p：告诉服务器将使用一个密码来登录，如果希望登录的用户名和密码为空，可以忽略此选项。

如登录本机的 MySQL 数据库，只需要输入以下命令：

```
mysql -u root -p
```

输入这个指令，回车后会提示输入密码，记住了上面第 3 步安装时的密码，填入即可登录成功，进入 MySQL 命令模式。

具体操作示例如下：

```
E:\mysql\mysqlinstall\mysql-8.0.11-winx64\bin>mysql -u root -p
Enter password: ************
Welcome to the MySQL monitor.  Commands end with ; or \g.
Your MySQL connection id is 8
Server version: 8.0.11

Copyright (c) 2000, 2018, Oracle and/or its affiliates. All rights reserved.

Oracle is a registered trademark of Oracle Corporation and/or its affiliates.
Other names may be trademarks of their respective owners.

Type 'help;' or '\h' for help. Type '\c' to clear the current input statement.

mysql>
```

由操作结果可以看到，可以进入 MySQL 命令模式了。

这里通过操作会发现，这个密码并不好记，如果每次登录都只能用这个密码，那会不会太麻烦了，可以更改密码吗？不要着急，接下来讲解如何更改这个密码。

在上面 MySQL 命令模式后面输入如下指令：

```
mysql>ALTER USER 'root'@'localhost' IDENTIFIED WITH mysql_native_password BY '新密码';
```

输入上面的指令，回车后就完成密码的更改了。当然，在执行上面指令前，需要想好自己想要的密码，并替换新密码三个字。

比如，想把新密码设置为 root，操作方式如下：

```
mysql> ALTER USER 'root'@'localhost' IDENTIFIED WITH mysql_native_password BY 'root';
Query OK, 0 rows affected (0.17 sec)
```

若执行操作后输出"Query OK"字符，则表明密码更改成功。

要查看更改后的密码是否生效，可以先退出 MySQL 的指令模式，再从 bin 目录下输入对应指令进入。

完成密码修改后，退出指令模式，再用新密码进入指令模式的具体操作如下：

```
E:\mysql\mysqlinstall\mysql-8.0.11-winx64\bin>mysql -u root -p
```

```
Enter password: ************
Welcome to the MySQL monitor.  Commands end with ; or \g.
Your MySQL connection id is 8
Server version: 8.0.11

Copyright (c) 2000, 2018, Oracle and/or its affiliates. All rights reserved.

Oracle is a registered trademark of Oracle Corporation and/or its
affiliates. Other names may be trademarks of their respective
owners.

Type 'help;' or '\h' for help. Type '\c' to clear the current input statement.

mysql> ALTER USER 'root'@'localhost' IDENTIFIED WITH mysql_native_password BY 'root';
Query OK, 0 rows affected (0.17 sec)

mysql> exit
Bye

E:\mysql\mysqlinstall\mysql-8.0.11-winx64\bin>mysql -u root -p
Enter password: ****
Welcome to the MySQL monitor.  Commands end with ; or \g.
Your MySQL connection id is 9
Server version: 8.0.11 MySQL Community Server - GPL

Copyright (c) 2000, 2018, Oracle and/or its affiliates. All rights reserved.

Oracle is a registered trademark of Oracle Corporation and/or its affiliates. Other names
may be trademarks of their respective owners.

Type 'help;' or '\h' for help. Type '\c' to clear the current input statement.

mysql>
```

执行以上操作以后，密码修改就完成了。

至此，Windows 下 MySQL 的安装和密码修改的操作都完成了。Linux 或 Mac 下，MySQL 的安装更简单，这里不再具体介绍。

若确实需要，Mac 下 MySQL 的安装可以参考以下博文地址写的图解安装方式：

https://blog.csdn.net/youzhouliu/article/details/80782892

Windows 下 MySQL 的安装博客文章链接如下：

https://blog.csdn.net/youzhouliu/article/details/80782125

【知识拓展：MySQL 的几种大版本】

① MySQL Community Server：社区版本，开源免费，但不提供官方技术支持。

② MySQL Enterprise Edition：企业版本，需付费，可以试用 30 天。

③ MySQL Cluster：集群版，开源免费，可将几个 MySQL Server 封装成一个 Server。

④ MySQL Cluster CGE：高级集群版，需付费。

⑤ MySQL Workbench（GUI TOOL）：专为 MySQL 设计的 ER/数据库建模工具，是著名的数据库设计工具 DBDesigner4 的继任者。MySQL Workbench 又分为两个版本，分别是社区版（MySQL Workbench OSS）、商用版（MySQL Workbench SE）。

MySQL Community Server 是开源免费的，这也是大家常用的 MySQL 版本。

另一方面，根据所使用操作系统平台的不同，MySQL 的各大版本又细分为多个更小的版本，如 Linux 版、Mac 版、Windows 版等。

2.3 MySQL 基本操作

2.3.1 MySQL 基本管理

进入 MySQL 的指令模式后，就可以开始 MySQL 的基本管理了。MySQL 的基本管理内容有查看数据库、选择数据库、查看数据表、查看数据表结构、查看数据表索引等。

1．查看数据库

进入 MySQL 的指令模式界面，输入 SHOW databases 指令，就会显示当前登录的 MySQL 下的所有数据库。操作及结果如下：

```
mysql> SHOW databases;
+--------------------+
| Database           |
+--------------------+
| information_schema |
| mysql              |
| performance_schema |
| sys                |
+--------------------+
4 rows in set (0.00 sec)
```

由上面输出结果可以看到，当前登录的 MySQL 系统中，Database（数据库）下面展示了四个数值，即表示当前数据库系统中有四个数据库。这四个库是默认创建的，更详细信息可以参考附录 A。

2．选择数据库

数据表是放置在数据库中的，需要先进入某个数据库。在终端进入某个数据库的指令语法为：

```
USE 数据库名
```

比如，要进入 MySQL 库，则操作如下：

```
mysql> USE mysql;
Database changed
```

3．查看数据表

进入数据库后，就可以查看数据库中的数据表，查看数据表的语法为：

```
SHOW tables;
```

如查看 MySQL 库中的表，则操作如下：

```
mysql> SHOW tables;
+---------------------------+
| Tables_in_mysql           |
+---------------------------+
| columns_priv              |
```

```
| component              |
| db                     |
| default_roles          |
| engine_cost            |
| func                   |
...
...
| time_zone_transition_type |
| user                   |
+---------------------------+
33 rows in set (0.00 sec)
```

4. 查看数据表结构

查看某个数据表的具体信息的命令如下：

```
SHOW columns FROM 数据表
```

该命令将显示数据表的属性、属性类型、主键信息、是否为 NULL、默认值等其他信息。具体操作示例如下：

```
mysql> SHOW columns FROM help_topic;
+------------------+---------------------+------+-----+---------+-------+
| Field            | Type                | Null | Key | Default | Extra |
+------------------+---------------------+------+-----+---------+-------+
| help_topic_id    | int(10) unsigned    | NO   | PRI | NULL    |       |
| name             | char(64)            | NO   | UNI | NULL    |       |
| help_category_id | smallint(5) unsigned| NO   |     | NULL    |       |
| description      | text                | NO   |     | NULL    |       |
| example          | text                | NO   |     | NULL    |       |
| url              | text                | NO   |     | NULL    |       |
+------------------+---------------------+------+-----+---------+-------+
6 rows in set (0.00 sec)
```

5. 查看数据表索引

显示数据表的索引信息的命令如下：

```
SHOW index FROM 数据表
```

该命令将显示数据表的详细索引信息，包括 PRIMARY KEY（主键）信息。具体操作示例如下：

```
mysql> SHOW index FROM help_topic;
+---------------+------------+----------+--------------+-------------+-----------
+---------------+------------+----------+--------------+-------------+-----------
+---------------+
| Table         | Non_unique | Key_name | Seq_in_index | Column_name | Collation
| Cardinality   | Sub_part   | Packed   | Null         | Index_type  | Comment
| Index_comment | Visible    |
+---------------+------------+----------+--------------+-------------+-----------
+---------------+------------+----------+--------------+-------------+-----------
+---------------+
| help_topic    | 0          |          | PRIMARY KEY  | 1           | help_topic_id | A
| 737           | NULL       | NULL     | BTREE        |             | YES
| help_topic    | 0          |          | name         | 1           | name        | A
```

```
| 642            | NULL        | NULL      |                | BTREE         |
|                | YES         |           |
+----------------+-------------+-----------+----------------+---------------+-----------
+----------------+-------------+-----------+----------------+---------------+-----------
+----------------+
2 rows in set (0.00 sec)
```

2.3.2 MySQL 数据库操作

MySQL 的数据库操作有创建数据库、删除数据库、选择数据库等操作。选择数据库已经在前面介绍，下面分别介绍创建数据库和删除数据库。

1．创建数据库

在登录 MySQL 服务后，可以在指令模式界面使用 CREATE 命令创建数据库，语法如下（不区分大小写）：

```
CREATE database 数据库名;
```

比如，创建一个名为 test 的数据库的示例代码如下：

```
mysql> CREATE database test;
```

具体示例及结果如下：

```
mysql> CREATE database test;
Query OK, 1 row affected (0.16 sec)

mysql> SHOW databases;
+--------------------+
| Database           |
+--------------------+
| information_schema |
| mysql              |
| performance_schema |
| sys                |
| test               |
+--------------------+
5 rows in set (0.00 sec)
```

由结果可知，数据库 test 创建成功了，通过 SHOW database 命令可以看到输出结果中新增了数据库 test。

2．删除数据库

数据库可以创建，也可以删除。删除数据库 DROP 命令的语法如下（不区分大小写）：

```
DROP database 数据库名;
```

比如，删除前面创建的 test 库的示例代码如下：

```
mysql> DROP DATABASE test;
```

具体示例及结果如下：

```
mysql> DROP database test;
Query OK, 0 rows affected (0.55 sec)
```

```
mysql> SHOW databases;
+--------------------+
| Database           |
+--------------------+
| information_schema |
| mysql              |
| performance_schema |
| sys                |
+--------------------+
4 rows in set (0.00 sec)
```

由结果可知，test 库被删除了，通过 SHOW database 命令可以看到输出结果中已经不存在名为 test 的数据库了。

在删除数据库过程中务必十分谨慎，因为在执行删除命令后，所有数据将消失。

2.3.3 MySQL 数据类型

MySQL 中定义数据字段的类型对数据库的优化是非常重要的。MySQL 支持多种类型，大致可以分为三类：数值、日期/时间和字符串（字符）类型。

1．数值类型

MySQL 支持所有标准的 SQL 数值数据类型，包括严格数值数据类型（INTEGER、SMALLINT、DECIMAL 和 NUMERIC），以及近似数值数据类型（FLOAT、REAL 和 DOUBLE PRECISION）。

关键字 INT 是 INTEGER 的同义词，关键字 DEC 是 DECIMAL 的同义词。

BIT 数据类型保存位字段值，并且支持 MyISAM、MEMORY、InnoDB 和 BDB 表。

作为 SQL 标准的扩展，MySQL 也支持整数类型 TINYINT、MEDIUMINT 和 BIGINT。表 2-1 显示了需要的每个整数类型的存储和范围。

2．日期和时间类型

表示时间值的日期和时间类型为 DATETIME、DATE、TIMESTAMP、TIME 和 YEAR。

每个时间类型有一个有效值范围和一个"零"值，当指定不合法的 MySQL 不能表示的值时，使用"零"值替代。

TIMESTAMP 类型有专有的自动更新特性。

日期和时间类型的各属性如表 2-2 所示。

3．字符串类型

字符串类型指 CHAR、VARCHAR、BINARY、VARBINARY、BLOB、TEXT、ENUM 和 SET。表 2-3 展示了字符串类型如何工作以及如何在查询中使用。

CHAR 和 VARCHAR 类型类似，但它们保存和检索的方式不同。它们的最大长度和尾部空格是否被保留等方面也不同。在存储或检索过程中不进行大小写转换。

BINARY 和 VARBINARY 类似 CHAR 和 VARCHAR，不同的是，它们包含二进制字符串而不是非二进制字符串。也就是说，它们包含字节字符串而不是字符字符串。它们没有字符集，排序和比较基于列值字节的数值。

表 2-1 整数类型的存储和范围

类 型	大 小	范围（有符号）	范围（无符号）	用 途
TINYINT	1字节	(-128, 127)	(0, 255)	小整数值
SMALLINT	2字节	(-3 2768, 3 2767)	(0, 65 535)	大整数值
MEDIUMINT	3字节	(-8 388 608, 8 388 607)	(0, 16 777 215)	大整数值
INT 或 INTEGER	4字节	(-2 147 483 648, 2 147 483 647)	(0, 4 294 967 295)	大整数值
BIGINT	8字节	(-9 233 372 036 854 775 808, 9 223 372 036 854 775 807)	(0, 18 446 744 073 709 551 615)	极大整数值
FLOAT	4字节	(-3.402 823 466 E+38, -1.175 494 351 E-38, 0, 1.175 494 351 E-38, 3.402 823 466 E+38)	0, (1.175 494 351 E-38, 3.402 823 466 E+38)	单精度浮点数值
DOUBLE	8字节	(-1.797 693 134 862 315 7 E+308, -2.225 073 858 507 201 4 E-308), 0, (2.225 073 858 507 201 4 E-308, 1.797 693 134 862 315 7 E+308)	0, (2.225 073 858 507 201 4 E-308, 1.797 693 134 862 315 7 E+308)	双精度浮点数值
DECIMAL	对DECIMAL(M,D)，若M>D，则为M+2，否则为D+2	依赖M和D的值	依赖M和D的值	小数值

表 2-2 日期和时间类型属性

类型	大小（字节）	范 围	格 式	用 途
DATE	3	1000-01-01/9999-12-31	YYYY-MM-DD	日期值
TIME	3	'-838:59:59'/'838:59:59'	HH:MM:SS	时间值或持续时间
YEAR	1	1901/2155	YYYY	年份值
DATETIME	8	1000-01-01 00:00:00/9999-12-31 23:59:59	YYYY-MM-DD HH:MM:SS	混合日期和时间值
TIMESTAMP	4	1970-01-01 00:00:00/2038 结束时间是第 2147483647 秒，北京时间 2038-1-19 11:14:07，格林尼治时间 2038 年 1 月 19 日凌晨 03:14:07	YYYYMMDD HHMMSS	混合日期和时间值，时间戳

表 2-3 字符串类型

类 型	大 小	用 途
CHAR	0～255 字节	定长字符串
VARCHAR	0～65535 字节	变长字符串
TINYBLOB	0～255 字节	不超过 255 个字符的二进制字符串
TINYTEXT	0～255 字节	短文本字符串
BLOB	0～65 535 字节	二进制形式的长文本数据
TEXT	0～65 535 字节	长文本数据
MEDIUMBLOB	0～16 777 215 字节	二进制形式的中等长度文本数据
MEDIUMTEXT	0～16 777 215 字节	中等长度文本数据
LONGBLOB	0～4 294 967 295 字节	二进制形式的极大文本数据
LONGTEXT	0～4 294 967 295 字节	极大文本数据

BLOB 是一个二进制大对象，可以容纳可变数量的数据，有 4 种：TINYBLOB、BLOB、MEDIUMBLOB 和 LONGBLOB，区别在于可容纳存储范围的不同。

TEXT 类型有 4 种：TINYTEXT、TEXT、MEDIUMTEXT 和 LONGTEXT。它们可存储的最大长度不同，可根据实际情况选择。

2.3.4　MySQL 数据表操作

数据表的基本操作包括：创建数据表、删除数据表、插入数据、查询数据、WHERE 子句、UPDATE 更新、DELETE 语句等。

开始下一步讲解之前，继续请出久违的"Python 快乐学习班"的同学。在《Python 实用教程》中的 Python 库一别之后，他们现在来到了数字校园。在数字校园中，他们将学习到什么新知识呢？现在一起去看看。

1．创建数据表

创建 MySQL 数据表需要以下信息：表名，表字段名，每个表的字段。

以下为创建 MySQL 数据表的 SQL 通用语法：

```
CREATE table table_name(column_name column_type);
```

比如，为方便记录"Python 快乐学习班"所有同学的基本信息，需要在 MySQL 中创建 data_school 库，从中创建一张名为 python_class 的数据表，表中有自增主键、学生学号、学生姓名、学生班级名称几个字段。这个操作的实现如下所示。

创建 data_school 数据库：

```
mysql> CREATE database data_school;
Query OK, 1 row affected (0.10 sec)
```

在 data_school 数据库中创建 python_class 表：

```
mysql> USE data_school;
Database changed
mysql> SHOW tables;
Empty set (0.00 sec)
```

```
mysql> CREATE TABLE IF NOT EXISTS 'python_class' (
    -> 'id' INT UNSIGNED AUTO_INCREMENT,
    -> 'number' INT(10) NOT NULL,
    -> 'name' VARCHAR(50) NOT NULL,
    -> 'class_name' VARCHAR(50) NOT NULL,
    -> PRIMARY KEY ('id')
    -> )ENGINE=InnoDB DEFAULT CHARSET=utf8;
Query OK, 0 rows affected, 1 warning (0.82 sec)

mysql> SHOW tables;
+----------------------+
| Tables_in_data_school |
+----------------------+
| python_class         |
+----------------------+
1 row in set (0.00 sec)
```

由执行结果看到,创建了一个名为 python_class 的表。

【示例解析】

如果不想字段为 NULL,可以设置字段的属性为 NOT NULL,在操作数据库时,如果输入该字段的数据为 NULL,就会报错。

AUTO_INCREMENT 定义列为自增的属性,一般用于主键,数值会自动加 1。

PRIMARY KEY 用于定义列为主键。可以使用多列来定义主键,列间以逗号分隔。

ENGINE 设置存储引擎,CHARSET 设置编码。

在实际项目应用中,创建数据表是一个比较常用的操作。不过,在创建一个数据表之前,需要先进行表结构的设计,否则会在创建表与删除表之间来回操作,或是需要经常更改表结构。

2. 删除数据表

数据表可以创建,也可以删除。MySQL 中数据表删除的通用语法如下:

```
DROP table table_name;
```

其中,table_name 指的是表名。

如要删除上面创建的 python_class 表,具体操作如下:

```
mysql> SHOW tables;
+----------------------+
| Tables_in_data_school |
+----------------------+
| python_class         |
+----------------------+
1 row in set (0.00 sec)

mysql> DROP tables python_class;
Query OK, 0 rows affected (0.32 sec)

mysql> SHOW tables;
Empty set (0.00 sec)
```

执行删除 python_class 表的操作后,可以看到数据库中 python_class 表已经不存在了。

MySQL中删除数据表是非常容易操作的,因为执行删除命令后所有数据都会消失,所以在进行删除表操作时要非常小心。

在实际项目应用中,删除表的操作是一个比较基本的操作,也是一个比较危险的操作。在没有经过足够的思考前,不要轻易进行表删除操作,否则会追悔莫及。

3. 插入数据

MySQL表中使用INSERT INTO SQL语句来插入数据。

MySQL数据表插入数据通用的INSERT INTO的语法如下:

```
INSERT INTO table_name(field1, field2,…,fieldN)
VALUES(value1, value2,…,valueN);
```

如果数据是字符型的,必须使用单引号或者双引号,如"小智"。

如在data_school中重新创建python_class表,向表中插入小萌、小智、小强三位同学的基本信息。具体操作如下:

```
mysql> INSERT INTO python_class
    -> (number, name, class_name)
    -> VALUES
    -> (1001, "小萌", "Python快乐学习班");
Query OK, 1 row affected (0.11 sec)

mysql> INSERT INTO python_class (number, name, class_name) VALUES (1002, "小智", "Python快乐学习班");
Query OK, 1 row affected (0.09 sec)

mysql> INSERT INTO python_class (number, name, class_name) VALUES (1003, "小强", "Python快乐学习班");
Query OK, 1 row affected (0.12 sec)
```

注意:箭头标记(->)不是SQL语句的一部分,仅仅表示一个新行。如果一条SQL语句太长,可以通过回车键创建一个新行来编写SQL语句,SQL语句的命令结束符为分号(;)。熟悉后,可以将一整条插入语句写在一行,如上面示例中的后面两条插入语句。

在以上实例中,并没有提供字段id的数据。因为在创建表的时候,字段id已经设置为AUTO_INCREMENT(自增)属性,所以,id字段会自增而不需要去设置。

以上示例中插入语句返回的结果是"Query OK, 1 row affected",这表明插入成功,但插入的结果形式该怎么查看呢,接下来看查询数据是怎样操作的。

在实际项目应用中,插入数据是一个非常基本的操作。插入数据的操作有不少技巧,特别在对插入效率有要求的操作中,可以有很多优化方式。在实际应用中遇到时,可以尝试做优化。

4. 查询数据

MySQL数据库使用SELECT语句来查询数据。MySQL数据库中查询数据通用的SELECT语法如下:

```
SELECT column_name, column_name FROM table_name;
```

SELECT命令可以读取一条或者多条记录。column_name为需要查询的字段名,使用column_name时,指定了column_name的字段才会被查询。也可以使用"*"来代替字段名,使用"*"时,SELECT语句会返回表的所有字段数据。

如查看前面插入的数据结果，操作如下：

```
mysql> SELECT * FROM python_class;
+----+--------+------+------------------+
| id | number | name | class_name       |
+----+--------+------+------------------+
| 1  | 1001   | 小萌 | Python快乐学习班 |
| 2  | 1002   | 小智 | Python快乐学习班 |
| 3  | 1003   | 小强 | Python快乐学习班 |
+----+--------+------+------------------+
3 rows in set (0.00 sec)

mysql> SELECT number,name  FROM python_class;
+--------+------+
| number | name |
+--------+------+
| 1001   | 小萌 |
| 1002   | 小智 |
| 1003   | 小强 |
+--------+------+
3 rows in set (0.00 sec)
```

由结果就能看到前面插入的数据了。

在大多数实际项目应用中，查询操作几乎是被执行最多的一个操作。所以，查询操作语句的编写更需要注意执行效率的问题，在越庞大复杂的系统中，对查询语句的要求会越高。在平时的学习积累中，我们需要更加注意编写高效率的查询语句。

5. WHERE 子句

MySQL 表中使用 SELECT 语句来读取数据，如果需有条件地从表中选取数据，可将 WHERE 子句添加到 SELECT 语句中，其语法如下：

```
SELECT field1, field2,…,fieldN  FROM table_name1, table_name2...
[WHERE condition1 [AND [OR]] condition2.....
```

查询语句中可以使用一个或者多个表，表之间使用逗号（,）分隔，并使用 WHERE 语句来设定查询条件，可以在 WHERE 子句中指定任何条件，可以使用 AND 或者 OR 指定一个或多个条件。

WHERE 子句类似程序语言中的 if 条件，根据表中的字段值来读取指定的数据。

WHERE 子句支持多种操作符，如表 2-4 所示（假定 A 为 10，B 为 20）：

表 2-4　WHERE 子句操作符

操作符	描　　述	实　例
=	等号，检测两个值是否相等，如果相等，则返回 true	(A = B) 返回 false
<>, !=	不等于号，检测两个值是否相等，如果不相等，则返回 true	(A != B) 返回 true
>	大于号，检测左边的值是否大于右边的值，如果左边的值大于右边的值，则返回 true	(A > B) 返回 false
<	小于号，检测左边的值是否小于右边的值，如果左边的值小于右边的值，则返回 true	(A < B) 返回 true
>=	大于等于号，检测左边的值是否大于或等于右边的值，如果左边的值大于或等于右边的值，则返回 true	(A >= B) 返回 false
<=	小于等于号，检测左边的值是否小于或等于右边的值，如果左边的值小于或等于右边的值，则返回 true	(A <= B) 返回 true

如果想在 MySQL 数据表中读取指定的数据，WHERE 子句是非常有用的。
使用主键作为 WHERE 子句的条件查询是非常快速的。
如果给定的条件在表中没有任何匹配的记录，那么查询不会返回任何数据。
如从 python_class 表中查找 name 为"小智"的记录，操作如下：

```
mysql> SELECT * FROM python_class WHERE name='小智';
+----+--------+------+----------------+
| id | number | name | class_name     |
+----+--------+------+----------------+
|  2 |  1002  | 小智 | Python 快乐学习班 |
+----+--------+------+----------------+
1 row in set (0.00 sec)
```

查找 id 为 1 的记录，操作如下：

```
mysql> SELECT * FROM python_class WHERE id=1;
+----+--------+------+----------------+
| id | number | name | class_name     |
+----+--------+------+----------------+
|  1 |  1001  | 小萌 | Python 快乐学习班 |
+----+--------+------+----------------+
1 row in set (0.00 sec)
```

查找 id 为 20 的记录，操作如下：

```
mysql> SELECT * FROM python_class WHERE id=20;
Empty set (0.00 sec)
```

由结果可知，python_class 表中没有 id 为 20 的记录，返回结果为空。
查找 number 为 1003 且 class_name 为 "Python 快乐学习班" 的记录，操作如下：

```
mysql> SELECT * FROM python_class WHERE number=1003 AND class_name='Python 快乐学习班';
+----+--------+------+----------------+
| id | number | name | class_name     |
+----+--------+------+----------------+
|  3 |  1003  | 小强 | Python 快乐学习班 |
+----+--------+------+----------------+
1 row in set (0.00 sec)
```

查找 number 为 1001 或 name 为 "小强" 的记录，操作如下：

```
mysql> SELECT * FROM python_class WHERE number=1001 or name='小强';
+----+--------+------+----------------+
| id | number | name | class_name     |
+----+--------+------+----------------+
|  1 |  1001  | 小萌 | Python 快乐学习班 |
|  3 |  1003  | 小强 | Python 快乐学习班 |
+----+--------+------+----------------+
2 rows in set (0.00 sec)
```

在实际项目应用中，查询语句的 WHERE 子句几乎是一个标配。对于查询语句，在试验或测试条件下不加 WHERE 子句是比较正常的。但是在实际项目中，若编写的查询语句没有使用 WHERE 子句做一些条件限制，基本上会存在一些性能问题，需要加以留意。

6. UPDATE 子句

修改或更新 MySQL 中的数据可以使用 UPDATE 命令来操作，其通用 SQL 语法如下：

```
UPDATE table_name SET field1=new-value1, field2=new-value2 [WHERE clause]
```

UPDATE 命令可以同时更新一个或多个字段，可以在 WHERE 子句中指定任何条件，也可以在一个单独表中同时更新数据。

当需要更新数据表中指定行的数据时，WHERE 子句是非常有用的。

若将 python_class 表中的 class_name 字段值更改为"Python 学习班"，则操作如下：

```
mysql> UPDATE python_class SET class_name='Python 学习班';
Query OK, 3 rows affected (0.16 sec)
Rows matched: 3  Changed: 3  Warnings: 0

mysql> SELECT * FROM python_class;
+----+--------+------+--------------+
| id | number | name | class_name   |
+----+--------+------+--------------+
|  1 |  1001  | 小萌 | Python 学习班 |
|  2 |  1002  | 小智 | Python 学习班 |
|  3 |  1003  | 小强 | Python 学习班 |
+----+--------+------+--------------+
3 rows in set (0.00 sec)
```

由结果看到，class_name 字段值都更改为了"Python 学习班"。

若需要将 name 为"小强"的记录的 number 字段值更改为 1005，操作如下：

```
mysql> UPDATE python_class SET number=1005 where name='小强';
Query OK, 1 row affected (0.18 sec)
Rows matched: 1  Changed: 1  Warnings: 0

mysql> SELECT * FROM python_class where name='小强';
+----+--------+------+--------------+
| id | number | name | class_name   |
+----+--------+------+--------------+
|  3 |  1005  | 小强 | Python 学习班 |
+----+--------+------+--------------+
1 row in set (0.00 sec)
```

由结果可知，name 为"小强"的记录的 number 字段值更改为了 1005。

UPDATE 命令的 WHERE 子句中可搭配各种条件，大家可以尝试。为后续讲解演示，将 class_name 的字段值更改回"Python 快乐学习班"。

在实际项目应用中，UPDATE 子句经常用于指定范围数据的更新。UPDATE 子句一般会使用 WHERE 子句进行条件限制。

7. DELETE 子句

DELETE 命令可以删除 MySQL 数据表中的记录，其通用语法如下：

```
DELETE FROM table_name [WHERE Clause]
```

如果没有指定 WHERE 子句，MySQL 表中的所有记录将被删除。

可以在 WHERE 子句中指定任何条件，也可以在单个表中一次性删除记录。

删除数据表中指定的记录时，WHERE 子句是非常有用的。

如删除 number 值为 1005 的记录，操作如下：

```
mysql> DELETE FROM python_class WHERE number=1005;
Query OK, 1 row affected (0.25 sec)


mysql> SELECT * FROM python_class;
+----+--------+--------+------------------+
| id | number | name   | class_name       |
+----+--------+--------+------------------+
|  1 |  1001  | 小萌   | Python 快乐学习班 |
|  2 |  1002  | 小智   | Python 快乐学习班 |
+----+--------+--------+------------------+
2 rows in set (0.00 sec)
```

由结果可知，number 值为 1005 的记录被删除了。

在实际项目应用中，DELETE 子句一般用于删除指定范围的记录，通常需要使用 WHERE 子句进行范围筛选。使用 DELETE 子句也需要小心，确定选择范围无误后再执行，以避免不必要的损失。

2.4 MySQL 高级操作

本节将讲解 MySQL 的高级操作，这些操作在实际项目应用中比较多，主要有 LIKE 子句、排序、分组等。

1. LIKE 子句

由前面的讲解知道，在 MySQL 中可以使用 SELECT 命令来读取数据，同时可以在 SELECT 语句中使用 WHERE 子句来获取指定的记录，在 WHERE 子句中可以使用等号（=）来设定获取数据的条件。

但是有时需要获取某字段中含有某个字符的所有记录，这时需要在 WHERE 子句中使用 LIKE 子句。

SELECT 语句使用 LIKE 子句从数据表中读取数据的通用语法如下：

```
SELECT field1, field2,...,fieldN
FROM table_name
WHERE field1 LIKE condition1 [AND [OR]] filed2 = xxxx
```

LIKE 子句中使用 "%" 字符来表示任意字符，如果没有使用 "%"，那么 LIKE 子句与 "=" 的效果相同。

可以在 WHERE 子句中指定任何条件，也可以在 WHERE 子句中使用 LIKE 子句，还可以使用 LIKE 子句代替 "="。

LIKE 通常与 "%" 一同使用，类似一个元字符的搜索。

可以使用 AND 或者 OR 指定一个或多个条件。

可以在 DELETE 或 UPDATE 命令中使用 WHERE…LIKE 子句来指定条件。

如查找 python_class 表中 name 字段值包含 "智" 字的记录，操作如下：

```
mysql> SELECT *  FROM python_class  WHERE name LIKE '%智%';
```

```
+----+--------+------+-------------------+
| id | number | name | class_name        |
+----+--------+------+-------------------+
| 2  | 1002   | 小智 | Python 快乐学习班 |
+----+--------+------+-------------------+
1 row in set (0.00 sec)
```

查找 python_class 表中 number 字段值以 03 结尾的记录，操作如下：

```
mysql> SELECT * FROM python_class WHERE number LIKE '%03';
+----+--------+------+-------------------+
| id | number | name | class_name        |
+----+--------+------+-------------------+
| 4  | 1003   | 小强 | Python 快乐学习班 |
+----+--------+------+-------------------+
1 row in set (0.00 sec)
```

由查找结果看到，找到了 number 字段值为 1003 的记录，%03 即匹配任意以 03 结尾的记录。

查找 python_class 表中 number 字段值以 10 开头的记录，操作如下：

```
mysql> SELECT * FROM python_class WHERE number LIKE '10%';
+----+--------+------+-------------------+
| id | number | name | class_name        |
+----+--------+------+-------------------+
| 1  | 1001   | 小萌 | Python 快乐学习班 |
| 2  | 1002   | 小智 | Python 快乐学习班 |
| 4  | 1003   | 小强 | Python 快乐学习班 |
+----+--------+------+-------------------+
3 rows in set (0.00 sec)
```

由操作结果看到，所有 number 字段值以 10 开头的记录都查找出来了。

LIKE 子句也称为模糊匹配，分为左模糊匹配（形如'a%'）、右模糊匹配（形如"%a'）和全模糊匹配（形如"%a%'）。使用全模糊匹配需要注意使用不上索引，查询效率会非常低下，在实际项目中涉及全模糊查询时，尽量转换为其他查询方式进行。

2. 排序

从 MySQL 中查找记录时，如果需要对读取的数据进行排序，可以使用 MySQL 的 ORDER BY 子句来设定按哪个字段哪种方式来进行排序，再返回搜索结果。

SELECT 语句使用 ORDER BY 子句将查询数据排序后再返回数据的通用语法如下：

```
SELECT field1, field2,…, fieldN table_name1, table_name2, …
ORDER BY field1, [field2, …] [ASC [DESC]]
```

MySQL 可以使用任何字段作为排序的条件，从而返回排序后的查询结果，也可以设定多个字段来排序，还可以添加 WHERE…LIKE 子句来设置查询条件。

可以使用 ASC 或 DESC 关键字来设置查询结果是按升序或降序排列。MySQL 中对查询结果默认按升序排列。

如将 python_class 表中记录按 number 字段升序排序，操作如下：

```
mysql> SELECT * FROM python_class ORDER BY number ASC;
```

```
+----+--------+------+---------------------+
| id | number | name | class_name          |
+----+--------+------+---------------------+
|  1 |   1001 | 小萌 | Python 快乐学习班   |
|  2 |   1002 | 小智 | Python 快乐学习班   |
|  4 |   1003 | 小强 | Python 快乐学习班   |
+----+--------+------+---------------------+
3 rows in set (0.00 sec)
```

或不需要指定排序方式，操作如下：

```
mysql> SELECT * FROM python_class ORDER BY number;
+----+--------+------+---------------------+
| id | number | name | class_name          |
+----+--------+------+---------------------+
|  1 |   1001 | 小萌 | Python 快乐学习班   |
|  2 |   1002 | 小智 | Python 快乐学习班   |
|  4 |   1003 | 小强 | Python 快乐学习班   |
+----+--------+------+---------------------+
3 rows in set (0.00 sec)
```

如将 python_class 表中记录按 number 字段降序排序，操作如下：

```
mysql> SELECT * FROM python_class ORDER BY number DESC;
+----+--------+------+---------------------+
| id | number | name | class_name          |
+----+--------+------+---------------------+
|  4 |   1003 | 小强 | Python 快乐学习班   |
|  2 |   1002 | 小智 | Python 快乐学习班   |
|  1 |   1001 | 小萌 | Python 快乐学习班   |
+----+--------+------+---------------------+
3 rows in set (0.00 sec)
```

在实际项目应用中，查询结果排序是比较常见的操作，除了对单个字段的排序，也可以对多个字段进行排序，多个字段的排序会根据字段次序优先进行排序。

3. 分组

GROUP BY 语句根据一列或多列对结果集进行分组。GROUP BY 语法如下：

```
SELECT column_name, function(column_name) FROM table_name
[WHERE column_name operator value]
GROUP BY column_name;
```

其中，function 表示 MySQL 中的一些函数，如 COUNT、SUM、AVG 等函数。

如对 python_class 表中的记录根据 class_name 分组，并统计个数，操作如下：

```
mysql> SELECT class_name,count(class_name) FROM python_class GROUP BY class_name;
+-------------------+-------------------+
| class_name        | count(class_name) |
+-------------------+-------------------+
| Python 快乐学习班 |                 3 |
+-------------------+-------------------+
1 row in set (0.00 sec)
```

GROUP BY 语句可用于数据的分组、去重等。

MySQL 的高级操作还有非常多，如 ALTER 命令、索引、连接、复制表、正则表达式、导入数据、导出数据等。这些操作在后续的项目应用中，用到对应的应用场景时会进行讲解，在本章就不展开做具体的讲解。

在实际项目应用中，分组操作是一个比较常见的操作。数据量比较大时，对同一个分组操作需求，根据写法不同，执行的效率相差会很大，在实际应用时，要多尝试，避免写低效率的语句。

2.5　小结

本章主要讲解了 MySQL 的一些基本知识，及部分在实际应用中使用频率比较高的高级操作。

本书不是主讲数据库的，所以在数据库深度这一块不会用更多的篇幅进行讲解。有需要更加深入 MySQL 的读者可以自行查看 MySQL 相关资料或书籍。

2.6　实战演练

1. 根据本章讲解内容，完成 MySQL 安装。
2. 在安装的 MySQL 上创建一个数据库，并在新建的数据库中创建数据表，库名及表名由自己定义。
3. 往创建的表中插入若干条记录，对表记录进行基本的增、删、改、查操作。
4. 再新建一张表，对当前创建的两张表进行关联查询。
5. 对上面创建的两张表进行修改，使两张表之间存在一定关联关系，实现将一张表的某些字段值添加到另一张表。
6. 再增加一张表，使新增的表与前面两张表有一定关系，在新表中增加若干记录。对当前三张表操作，找出三张表中的关联数据。

第 3 章 PyMySQL 的安装和操作

第 2 章介绍了 MySQL 的基本操作,但没有引入 Python 操作 MySQL 的方式。本章将介绍 Python 如何通过 PyMySQL 连接关系型数据库,以 MySQL 作为关系型数据库进行操作示例讲解。

Python 通过 PyMySQL 操作 MySQL 类似学校通过工作人员安排 IT 大讲堂听讲人员的座次。

对于进入 IT 大讲堂听讲的人员座次安排,工作人员可以根据学校要求更换某排人员的位置,也可以替换某个座位的人员,这就如同对 MySQL 的更改操作,工作人员类似 PyMySQL,学校则相当于 Python。工作人员也可以将某排所有听众先撤空,类似 MySQL 中的删除操作,不影响其他排,也保持排列顺序。下面各节将进行更细致的讲解。

3.1 PyMySQL 的介绍与安装

为了使 Python 连接数据库,需要一个驱动,这个驱动是用于与数据库交互的库。在 Python 3.x 版本中,PyMySQL 是从 Python 连接到 MySQL 数据库服务器的接口,在 Python 2 中则使用 MySQLDB。PyMySQL 的目标是成为 MySQLDB 的替代品。

PyMySQL 是一个开源项目,支持如下 Python 版本:Python 2,Python 2.7,Python 3 及以上。PyMySQL 遵循 Python 数据库 API v2.0 规范,包含 pure-Python MySQL 客户端库。

在使用 PyMySQL 前,需要确保计算机上安装了 PyMySQL。如果没有安装 PyMySQL,在 Windows、Linux 或 Mac 系统下,都可以通过如下命令安装(使用 pip 或 pip3):

```
pip install PyMySQL
```

怎么检查 PyMySQL 是否安装成功?检查比较简单,如在 Windows 系统中,可以按如下操作进行检查。打开命令提示符:

```
C:\Users\lyz>python
Python 3.6.5 (v3.6.5:f59c0932b4, Mar 28 2018, 17:00:18) [MSC v.1900 64 bit (AMD64)] on win32
Type "help", "copyright", "credits" or "license" for more information.
>>>
```

然后输入如下命令:

```
>>> import pymysql
>>>
```

若安装成功,则输入"import pymysql"命令后,光标会定位到下一行命令提示符,否则会提示如下错误信息:

```
>>> import pymysql
Traceback (most recent call last):
  File "<stdin>", line 1, in <module>
```

```
ModuleNotFoundError: No module named 'pymysql'
>>>
```

若在执行 import 语句时出现"ModuleNotFoundError: No module named 'pymysql'"这样的提示,则表示 pymysql 模块尚未安装,使用上面的安装语句进行安装即可。注意:使用 pip 安装模块时,可能需要管理员或 root 权限,安装时根据提示正确执行即可。

3.2 PyMySQL 连接 MySQL 数据库

PyMySQL 安装成功后,就可以连接 MySQL 数据库了。在连接之前有两个概念要先理解:连接对象和游标对象。

连接(Connect)对象:

```
class pymysql.connections.Connection(host=None, user=None, password='',
database=None, port=0, unix_socket=None, charset='', sql_mode=None, read_default_file=None,
conv=None, use_unicode=None, client_flag=0, cursorclass=<class 'pymysql.cursors.Cursor'>,
init_command=None, connect_timeout=10, ssl=None, read_default_group=None, compress=None,
named_pipe=None, autocommit=False, db=None, passwd=None, local_infile=False,
max_allowed_packet=16777216, defer_connect=False, auth_plugin_map=None, read_timeout=None,
write_timeout=None, bind_address=None, binary_prefix=False, program_name=None,
server_public_key=None)
```

用 MySQL 服务器表示套接字。

获取此类实例的正确方法是调用 connect()方法建立与 MySQL 数据库的连接。

连接对象中几个关键参数的解释如下(全部参数的解释参见附录 B)。

❖ host:数据库服务器所在的主机。
❖ user:以登录身份登录的用户名。
❖ password:要使用的密码。
❖ database:要使用的数据库,设置为 None,则指不使用特定的数据库。
❖ port:要使用的 MySQL 端口,默认即可(默认值为 3306)。

获取数据库连接的基本语法如下:

```
pymysql.connect(host, user, password, database, port)
```

其中的参数对应上面参数解释中的值。

一般使用连接对象时,还会使用连接对象的如下方法。

❖ close():发送退出消息并关闭套接字。
❖ commit():提交更改到稳定存储。
❖ cursor(cursor=None):创建一个新游标以执行查询。cursor 参数指要创建的游标类型,即 Cursor、SSCursor、DictCursor、SSDictCursor 之一,None 指使用 Cursor。
❖ rollback():回滚当前事务。

这里大概介绍,后面示例中使用时会有更详尽的描述。

游标(Cursor)对象:

```
class pymysql.cursors.Cursor(connection)
```

这是用于与数据库交互的对象。

不要自己创建 Cursor 实例,调用 connections.Connection.cursor()即可。

游标对象的方法如下。
- close()：关闭光标，会释放所有剩余数据。
- execute(query, args=None)：执行查询。其中，query（字符型）参数为需要执行的查询。args（元组、列表或字典类型）为与查询一起使用的参数（可选）。返回受影响的行数（如果有），返回数据的类型为 INT。
- fetchone()：获取一行。
- fetchall()：获取所有行。
- fetchmany(size=None)：获取指定的 size 行。

这里大概介绍，后面示例中使用时会有更详尽的描述。

下面是使用 PyMySQL 连接 MySQL 数据库的示例（mysql_conn_exp.py）：

```python
import pymysql

# 打开数据库连接，不加端口号写法
db = pymysql.connect("localhost", "root", "root", "data_school")
# 使用数据库连接对象的 cursor()方法创建一个游标对象 cursor
cursor = db.cursor()
# 使用游标对象的 execute()方法执行 SQL 查询
cursor.execute("SELECT VERSION()")
# 使用游标对象的 fetchone()方法获取单条数据.
data = cursor.fetchone()
print(f"Database version:{data}")
# 关闭数据库连接
db.close()
```

执行程序，得到执行结果如下：

```
Database version:('8.0.11',)
```

对 mysql_conn_exp.py 中代码的解释如下。

代码 import pymysql：导入 pymysql 库。

代码 db=pymysql.connect("localhost", "root", "root", "data_school")：打开数据库连接，参数值对应如下：

- host 赋值为 localhost，因为连接的是本地，非本地连接要填写对应的 IP 地址。
- user 赋值为 root，用户名为 root。
- password 赋值为 root，这是一个权限最高的用户，实际应用中要慎用这么高权限的用户名和密码。
- database 赋值为 data_school，是在第 2 章中已经创建好的一个数据库，这里直接拿来使用。

该行代码也可以写成如下形式：

```python
# 打开数据库连接，添加端口号写法
db = pymysql.connect("localhost", "root", "root", "data_school", 3306)
```

或写成如下形式，效果也是一样的。

```python
# 打开数据库连接，显示指明参数名写法
db = pymysql.connect(host="localhost", user="root", password="root", database="data_school", port=3306)
```

代码 cursor = db.cursor()：用数据库连接对象的 cursor()方法创建一个游标对象 cursor。

代码 cursor.execute("SELECT VERSION()")：用游标对象的 execute()方法执行查询。
代码 data = cursor.fetchone()：用游标对象的 fetchone()方法获取单条数据。
代码 db.close()：关闭数据库连接。
这里是对 MySQL 数据库连接的简单介绍，接下来展示对 MySQL 数据库的更多操作。

3.3 PyMySQL 对 MySQL 数据库的基本操作

本节将介绍通过 PyMySQL 对 MySQL 数据库的增加、查询、更改、删除等操作，在第 2 章创建的 python_class 表上进行。本章所有代码存放在源码目录的 chapter3 目录下。

1. 数据库插入操作

现在需要通过编写 Python 代码把"小强"信息插入到 python_class 表中，"小强"信息为：number 为 1005，name 为小强，class_name 为 Python 快乐学习班。

实现示例代码如下（insert_exp_01.py）：

```python
import pymysql

def insert_record():
    """
    插入数据
    :return:
    """
    # 打开数据库连接，添加端口号写法
    db = pymysql.connect("localhost", "root", "root", "data_school", 3306)
    # 使用 cursor()方法获取操作游标
    cursor = db.cursor()

    # SQL 插入语句
    sql = "INSERT INTO python_class(number,name, class_name) \
        VALUES ({}, '{}', '{}')".format(1005, '小强', 'Python 快乐学习班')
    try:
        # 执行 SQL 语句
        cursor.execute(sql)
        # 执行 SQL 语句
        db.commit()
    except:
        # 发生错误时回滚
        db.rollback()

    # 关闭数据库连接
    db.close()

if __name__ == "__main__":
    insert_record()
```

执行 insert_exp_01.py 文件前先查看 python_class 表中的数据情况：

```
mysql> SELECT * FROM python_class;
+----+--------+------+-----------------+
| id | number | name | class_name      |
```

```
+----+--------+------+----------------------+
| 1  |  1001  | 小萌 | Python 快乐学习班    |
| 2  |  1002  | 小智 | Python 快乐学习班    |
| 4  |  1003  | 小强 | Python 快乐学习班    |
+----+--------+------+----------------------+
3 rows in set (0.00 sec)
```

执行 insert_exp_01.py 文件后,python_class 表中的数据情况如下:

```
mysql> SELECT * FROM python_class;
+----+--------+------+----------------------+
| id | number | name | class_name           |
+----+--------+------+----------------------+
| 1  |  1001  | 小萌 | Python 快乐学习班    |
| 2  |  1002  | 小智 | Python 快乐学习班    |
| 4  |  1003  | 小强 | Python 快乐学习班    |
| 5  |  1005  | 小张 | Python 快乐学习班    |
+----+--------+------+----------------------+
4 rows in set (0.00 sec)
```

由结果可知,小强的信息已成功插入 python_class 表中。

文件 insert_exp_01.py 的代码也可以写成 insert_exp_02.py 所示示例:

```python
import pymysql

def insert_record():
    """
    插入数据
    :return:
    """
    # 打开数据库连接,添加端口号写法
    db = pymysql.connect("localhost", "root", "root", "data_school", 3306)
    # 使用 cursor()方法获取操作游标
    cursor = db.cursor()
    # SQL 插入语句
    sql = """INSERT INTO python_class(number,name, class_name)
            VALUES (1005, '小张', 'Python 快乐学习班')"""
    try:
        # 执行 SQL 语句
        cursor.execute(sql)
        # 提交到数据库执行
        db.commit()
    except:
        # 如果发生错误则回滚
        db.rollback()

    # 关闭数据库连接
    db.close()

if __name__ == "__main__":
    insert_record()
```

该代码的执行效果和 insert_exp_01.py 文件中代码执行效果一致。

在该示例代码中,不要忘记写 db.commit()这行代码,对于 MySQL 的更改操作,需要显式做事务的提交操作，否则会导致数据库中数据未成功写入，即数据丢失。

2. 数据库查询操作

如查看 python_class 表中 number 为 1002 的学生的所有信息，通过 Python 实现的示例代码如下（select_exp.py）：

```python
import pymysql

def mysql_select():
    """
    数据查找
    :return:
    """
    # 打开数据库连接，添加端口号写法
    db = pymysql.connect("localhost", "root", "root", "data_school", 3306)
    # 使用 cursor()方法获取操作游标
    cursor = db.cursor()

    # 待查询的学号
    s_num = 1002
    # SQL 查询语句
    sql = "SELECT *  FROM python_class  WHERE number = {}".format(s_num)
    try:
        # 执行 SQL 语句
        cursor.execute(sql)
        # 获取所有记录列表
        results = cursor.fetchall()
        for row in results:
            num = row[1]
            name = row[2]
            cs_name = row[3]
            # 打印结果
            print(f"学号为{s_num}的详细信息为:" f"number={num},name={name},class_name={cs_name}")
    except:
        print("Error: unable to fetch data")

    # 关闭数据库连接
    db.close()

if __name__ == "__main__":
    mysql_select()
```

执行该示例代码，得到的结果如下：

```
学号为 1002 的详细信息为:number=1002,name=小智,class_name=Python 快乐学习班
```

3. 数据库更新操作

若小张改名为小李，需要你在 python_class 表中将他的 name 值更改为小李，其他信息不变。实现示例代码如下（update_exp.py）：

```python
import pymysql

def update_mysql():
    """
```

```python
    数据更新
    :return:
    """
    # 打开数据库连接，添加端口号写法
    db = pymysql.connect("localhost", "root", "root", "data_school", 3306)
    # 使用 cursor()方法获取操作游标
    cursor = db.cursor()

    # SQL 更新语句
    sql = "UPDATE python_class SET name = '{}' WHERE name = '{}'".format('小李', '小张')
    try:
        # 执行 SQL 语句
        cursor.execute(sql)
        # 提交到数据库执行
        db.commit()
    except:
        # 发生错误时回滚
        db.rollback()

    # 关闭数据库连接
    db.close()

if __name__ == "__main__":
    update_mysql()
```

执行以上代码，可以从 MySQL 命令控制台查看执行结果。当然，前面已经学习了数据库查询操作，可以结合 Python 的函数编写方式，将 update_exp.py 中的代码改写为如下新的形式（update_exp_01.py）：

```python
import pymysql

# 打开数据库连接，添加端口号写法
db = pymysql.connect("localhost", "root", "root", "data_school", 3306)
# 使用 cursor()方法获取操作游标
cursor = db.cursor()

def query_mysql(s_num):
    """
    根据条件查找数据
    :param s_num:
    :return:
    """
    # SQL 查询语句
    sql = "SELECT *  FROM python_class  WHERE number={}".format(s_num)
    try:
        # 执行 SQL 语句
        cursor.execute(sql)
        # 获取所有记录列表
        results = cursor.fetchall()
        for row in results:
            num = row[1]
            name = row[2]
            # 打印结果
```

```python
            print(f"学号为{s_num}的详细信息为:number={num},name={name}")
    except:
        print("Error: unable to fetch data")

def update_mysql():
    """
    数据更新
    :return:
    """
    # SQL 更新语句
    sql = "UPDATE python_class SET name = '{}' WHERE name='{}'".format('小李', '小张')
    try:
        # 执行 SQL 语句
        cursor.execute(sql)
        # 提交到数据库执行
        db.commit()
    except:
        # 发生错误时回滚
        db.rollback()

if __name__ == "__main__":
    number = 1005
    print("--------------更改之前---------------")
    query_mysql(number)
    update_mysql()
    print("--------------更改之后---------------")
    query_mysql(number)
# 关闭连接
    db.close()
```

执行 update_exp_01.py 文件中的代码，得到输出结果如下：

```
--------------更改之前---------------
学号为 1005 的详细信息为:number=1005,name=小张
--------------更改之后---------------
学号为 1005 的详细信息为:number=1005,name=小李
```

4．删除操作

通过 Python 代码也可以对 MySQL 数据库进行删除操作。现在需要将 python_class 表中 number 为 1005 的记录删除，示例代码如下（delete_exp.py）：

```python
import pymysql

# 打开数据库连接，添加端口号写法
db = pymysql.connect("localhost", "root", "root", "data_school", 3306)
# 使用 cursor()方法获取操作游标
cursor = db.cursor()

def query_mysql(s_num):
    """
    根据条件查找数据
    :param s_num:
    :return:
    """
```

```python
        # SQL 查询语句
        sql = "SELECT *  FROM python_class  WHERE number={}".format(s_num)
        try:
            # 执行 SQL 语句
            cursor.execute(sql)
            # 获取所有记录列表
            results = cursor.fetchall()
            if results is None or len(results) == 0:
                print(f'没有找到学号为{s_num}的信息。')

            for row in results:
                num = row[1]
                name = row[2]
                # 打印结果
                print(f"学号为{s_num}的详细信息为:number={num},name={name}")
        except:
            print("Error: unable to fetch data")

def delete_mysql(num):
    """
    根据指定条件做删除
    :param num:
    :return:
    """
    # SQL 删除语句
    sql = "DELETE FROM python_class  WHERE number={}".format(num)
    try:
        # 执行 SQL 语句
        cursor.execute(sql)
        # 提交修改
        db.commit()
    except:
        # 发生错误时回滚
        db.rollback()

if __name__ == "__main__":
    number = 1005
    print('------删除之前--------')
    query_mysql(number)
    delete_mysql(number)
    print('------删除之后--------')
    query_mysql(number)
    # 关闭连接
    db.close()
```

执行该示例代码,得到执行结果如下:

```
------删除之前--------
学号为 1005 的详细信息为:number=1005,name=小李
------删除之后--------
没有找到学号为 1005 的信息。
```

5. 执行事务

前面的示例代码中多处出现 db.rollback()这样的代码，含义是发生错误时回滚。为什么有些操作需要回滚，有些却不需要呢？这就涉及事务问题，而事务机制可以确保数据一致性。

事务具有 4 个属性：原子性（Atomicity）、一致性（Consistency）、隔离性（Isolation）、持久性（Durability）。这 4 个属性通常称为 ACID 特性，在第 1 章已经讲解。在对数据库表做操作时，对数据库表的插入、更新、删除操作都会涉及对数据的变更，为确保数据的一致性，一般这三个操作要在发生错误时做数据回滚，查询操作不会更改数据，不需要做数据的回滚。

对于支持事务的数据库，在 Python 数据库编程中，当游标建立时，就自动开始了一个隐形的数据库事务。

commit()方法中，游标的所有更新操作在遇到 rollback()方法时都会回滚当前游标的所有操作。每个 commit()方法都开始了一个新的事务。

6. 错误处理

同 Python 程序一样，通过 Python 代码操作 MySQL 数据库时会经常遇到不同数据库层的异常，DB API 中定义了一些数据库操作的异常，如表 3-1 所示。

表 3-1 数据库操作的异常

异 常	描 述
Warning	当有严重警告时触发，如插入数据时被截断等，必须是 StandardError 的子类
Error	警告以外所有其他错误类，必须是 StandardError 的子类
InterfaceError	当有数据库接口模块本身的错误（而不是数据库的错误）发生时触发，必须是 Error 的子类
DatabaseError	与数据库有关的错误发生时触发，必须是 Error 的子类
DataError	当有数据处理时的错误发生时触发，如除零错误、数据超范围等，必须是 DatabaseError 的子类
OperationalError	指非用户控制的而是操作数据库时发生的错误，如连接意外断开、数据库名未找到、事务处理失败、内存分配错误等操作数据库时发生的错误，必须是 DatabaseError 的子类
IntegrityError	完整性相关的错误，如外键检查失败等，必须是 DatabaseError 的子类
InternalError	数据库的内部错误，如游标（cursor）失效了、事务同步失败等，必须是 DatabaseError 子类
ProgrammingError	程序错误，如数据表（table）没找到或已存在、SQL 语句语法错误、参数数量错误等，必须是 DatabaseError 的子类
NotSupportedError	不支持错误，指使用了数据库不支持的函数或 API 等。例如在连接对象上使用.rollback()函数，然而数据库并不支持事务或者事务已关闭。必须是 DatabaseError 的子类

3.4 PyMySQL 操作多表

到目前为止，所有操作都是在单表上操作的，实际应用中，在一个 SQL 语句中操作多张表的操作是比较常见的。本节介绍 Python 代码通过 PyMySQL 操作 MySQL 中多张表的操作。

在 data_school 库中准备另一个表：学生地址表 st_addr，包含学生学号、学生家庭住址等信息。st_addr 表的创建通过 Python 代码实现，实现如下（create_table_exp.py）：

```
import pymysql
```

```python
def create_table():
    """
    创建表
    :return:
    """
    # 打开数据库连接，添加端口号写法
    db = pymysql.connect("localhost", "root", "root", "data_school", 3306)
    # 使用 cursor() 方法获取操作游标
    cursor = db.cursor()

    # 使用预处理语句创建表
    sql = """CREATE table st_addr (id INT UNSIGNED AUTO_INCREMENT,
                                    number INT(10) NOT NULL,
                                    addr VARCHAR(100) NOT NULL,
                                    PRIMARY KEY (id)
                                    )ENGINE=InnoDB DEFAULT CHARSET=utf8"""

    cursor.execute(sql)

    # 关闭数据库连接
    db.close()

if __name__ == "__main__":
    create_table()
```

执行代码后，通过指令面板查看 MySQL 数据库表，结果如下：

```
mysql> SHOW tables;
+----------------------+
| Tables_in_data_school |
+----------------------+
| python_class         |
| st_addr              |
+----------------------+
2 rows in set (0.01 sec)
```

由结果可知，data_school 中已经新增了一个名为 st_addr 的表。

创建 st_addr 表后，用批量插入的方式向表中插入几条数据，代码如下（query_exp.py）：

```python
import pymysql

def query_record():
    """
    记录查询
    :return:
    """
    # 打开数据库连接，不添加端口号写法
    db = pymysql.connect("localhost", "root", "root", "data_school")
    # 使用 cursor() 方法获取操作游标
    cursor = db.cursor()

    # SQL 查询语句
    sql = "SELECT * FROM st_addr"
    try:
```

```python
        # 执行 SQL 语句
        cursor.execute(sql)
        # 获取所有记录列表
        results = cursor.fetchall()
        for row in results:
            id_v = row[0]
            num = row[1]
            addr = row[2]
            # 打印结果
            print(f"详细信息为:id={id_v},number={num},addr={addr}")
    except:
        print("Error: unable to fetch data")

    # 关闭数据库连接
    db.close()

if __name__ == "__main__":
    query_record()
```

执行代码，得到结果如下：

```
详细信息为:id=1,number=1001,addr=A 城区
详细信息为:id=2,number=1002,addr=B 城区
详细信息为:id=3,number=1003,addr=C 城区
```

由执行结果可知，批量插入成功，数据查询也成功。

现在需要查找 number 为 1002 的同学的 name 和 addr，怎样能更方便查找呢？

这里需要引进 MySQL 连接的使用。"连接"概念在第 2 章有提到，但没有展开，这里对这个概念做一个补充。

MySQL 中一般通过 JOIN 在两个或多个表中查询数据，即一般通过 JOIN 做连接查询。按照功能，JOIN 大致分为如下三类：

① INNER JOIN（内连接，或等值连接，可以直接写成 JOIN）：获取两个表中字段匹配关系的记录。比如，A、B 两张表做内连接，查询到的记录是既在 A 表又在 B 表的记录，相当于数学中两个集合的交集。

② LEFT JOIN（左连接）：获取左表所有记录，即使右表没有对应匹配的记录。比如 A、B 两张表做左连接，A 为左表，B 为右表，左连接得到的查询结果是 A 表所有满足条件的记录都显示，B 表满足条件的显示对应结果，不满足条件的显示空，保持和 A 表查找到的记录条数一致。即以 A 表为标准，B 表不够的以空填补。

③ RIGHT JOIN（右连接）：与 LEFT JOIN 相反，用于获取右表所有记录，即使左表没有对应匹配的记录。比如，A、B 两张表做右连接，A 为左表，B 为右表，右连接得到的查询结果是 B 表所有满足条件的记录都显示，A 表满足条件的显示对应结果，不满足条件的显示空，保持与 B 表查找到的记录条数一致。即以 B 表为标准，A 表不够的以空填补。

若想知道更多，读者可以自行查找相关资料。

查找 number 为 1002 的同学的 name 和 addr 这个需求需要通过内连接来实现，在 MySQL 指令面板中的写法及结果如下：

```
mysql> SELECT a.number,a.name,b.addr  FROM python_class a join st_addr b on a.number=b.number
```

```
and a.number=1002;
+--------+------+-------+
| number | name | addr  |
+--------+------+-------+
|  1002  | 小智 | B城区 |
+--------+------+-------+
1 row in set (0.00 sec)
```

由结果可知，使用 JOIN 得到了想要的结果。这里需要注意后面的条件开始用的是 ON，不是 WHERE。

接下来看 Python 代码的实现方式，示例如下（mult_table_query_exp.py）：

```python
import pymysql

def query_mysql(num):
    """
    记录查询
    :return:
    """
    # 打开数据库连接，不添加端口号写法
    db = pymysql.connect("localhost", "root", "root", "data_school")
    # 使用 cursor()方法获取操作游标
    cursor = db.cursor()

    # SQL 查询语句
    sql = "SELECT a.number,a.name,b.addr FROM python_class a " \
          "JOIN st_addr b ON a.number=b.number AND a.number={}".format(num)

    try:
        # 执行 SQL 语句
        cursor.execute(sql)
        # 获取所有记录列表
        results = cursor.fetchall()
        for row in results:
            num_v = row[0]
            name = row[1]
            addr = row[2]
            # 打印结果
            print(f"学号为{num}的详细信息为:number={num_v},name={name},addr={addr}")
    except:
        print("Error: unable to fetch data")

    # 关闭数据库连接
    db.close()

if __name__ == "__main__":
    query_mysql(1002)
```

执行以上代码，得到执行结果如下：

```
学号为1002的详细信息为:number=1002,name=小智,addr=B城区
```

由结果可知，以上代码已实现多表的连接操作。

多表的连接操作还支持更改和删除等操作，此处不具体举例，大家可以自行尝试。

3.5 高级封装

从 3.3 节和 3.4 节的代码中可以看到,每个 PY 文件中都出现了不少需要重复编写的代码段,如下面两段代码,几乎每个 PY 文件中都要编写一遍:

```python
# 打开数据库连接,添加端口号写法
db = pymysql.connect("localhost", "root", "root", "data_school", 3306)
# 使用 cursor()方法获取操作游标
cursor = db.cursor()

# 关闭数据库连接
db.close()
```

那么,这些代码是否可以封装到一个 PY 文件中,从而在需要时直接调用呢?答案是可以的,现在在 chapter3 目录下创建一个名为 common 的目录,在其下创建一个名为 mysql_conn.py 的文件。在 mysql_conn.py 中按如下方式添加代码。创建一个名为 MySQLConnection 的类,代码如下:

```python
class MySQLConnection(object):
```

观察前面编写的打开数据库连接的语句,在 MySQLConnection 类中创建一个初始化方法,方法定义如下:

```python
def __init__(self, host=None, user=None, password=None, database=None, port=3306):
    self.host = host
    self.user = user
    self.password = password
    self.database = database
    self.port = port
```

其中,5 个参数分别对应 host、user、password、database 和 port。port 默认值为 3306,即 MySQL 连接的默认端口号。这 5 个参数对应 MySQL 数据库连接的 5 个基本参数,除了 port 参数,缺少其中任何一个参数都不能连接 MySQL。

这里定义了参数初始化方法,但只是参数,还需打开数据库连接。

在 MySQLConnection 类中增加如下代码:

```python
def get_db(self):
    """
    打开数据库连接
    :return:
    """
    db = pymysql.connect(self.host, self.user, self.password, self.database, self.port)
    return db
```

get_db()方法中实现了打开数据库连接的方式,并最终返回数据库连接对象。得到数据库连接对象后,还需要获取操作游标,但获取操作游标时需要数据库连接对象,所以需要在 __init__()方法中初始化数据库连接对象。__init__()方法更改如下:

```python
def __init__(self, host=None, user=None, password=None, database=None, port=3306):
    self.host = host
    self.user = user
    self.password = password
    self.database = database
```

```
    self.port = port
    self.db = self.get_db()
```

增加一行初始化获取数据库连接对象的代码。同时，在 MySQLConnection 类中增加如下方法：

```python
def conn(self):
    """
    使用 cursor()方法获取操作游标
    :return:
    """
    cursor = self.db.cursor()
    return cursor
```

由 conn()方法即可得到操作游标。得到操作游标后，需要在 MySQLConnection 类初始化时先初始化，在_init__()方法中增加一行代码。_init__()方法更改如下：

```python
def __init__(self, host=None, user=None, password=None, database=None, port=3306):
    self.host = host
    self.user = user
    self.password = password
    self.database = database
    self.port = port
    self.db = self.get_db()
    self.conn = self.conn
```

至此，MySQLConnection 类的初始化就完成了。

当然，在实际应用中，close()方法也是重复编写比较多的，把 close()方法也封装到类 MySQLConnection 中。在 MySQLConnection 类中添加 close()方法的代码如下：

```python
def close(self):
    """
    关闭数据库连接
    :return:
    """
    self.db.close()
```

这些代码添加完成后，MySQLConnection 类的代码结构如下：

```python
import pymysql

class MySQLConnection(object):
    """
    MySQL 连接
    """
    def __init__(self, host=None, user=None, password=None, database=None, port=3306):
        self.host = host
        self.user = user
        self.password = password
        self.database = database
        self.port = port
        self.db = self.get_db()
        self.conn = self.conn
```

```python
def get_db(self):
    """
    打开数据库连接
    :return:
    """
    db = pymysql.connect(self.host, self.user, self.password, self.database, self.port)
    return db

def conn(self):
    """
    使用 cursor()方法获取操作游标
    :return:
    """
    cursor = self.db.cursor()
    return cursor

def close(self):
    """
    关闭数据库连接
    :return:
    """
    self.db.close()
```

这种形式封装好的代码可以实现数据库连接的复用，还可以进一步封装，可以把数据查询和数据更新的方法也加以封装，并在封装的查询和更新方法中关闭数据库连接，从而不需在外部调用时关心数据库连接是否关闭的问题。

数据查询分为全量查询和查询一条数据。全量查询代码封装如下：

```python
def query_all(self, query_sql):
    """
    根据查询语句查询所有数据
    :param query_sql:
    :return:
    """
    try:
        cursor = self.conn()
        # 执行 SQL 语句
        cursor.execute(query_sql)
        # 获取所有记录列表
        results = cursor.fetchall()
        return results
    except Exception as ex:
        print("query all Error: {}".format(ex))
    finally:
        self.close()
```

查询一条数据代码封装如下：

```python
def query_one(self, query_sql):
    """
```

```python
    根据查询语句查询一条语句
    :param query_sql:
    :return:
    """
    try:
        cursor = self.conn()
        # 执行 SQL 语句
        cursor.execute(query_sql)
        # 获取一条记录
        results = cursor.fetchone()
        return results
    except Exception as ex:
        print("query all Error: {}".format(ex))
    finally:
        self.close()
```

数据更新代码封装如下：

```python
def update(self, update_sql):
    """
    根据更新语句更新数据
    :param update_sql:
    :return:
    """
    try:
        cursor = self.conn()
        # 执行 SQL 语句
        cursor.execute(update_sql)
        # 提交到数据库执行
        self.db.commit()
    except Exception as ex:
        print("update Error: {}".format(ex))
        # 发生错误时回滚
        self.db.rollback()
    finally:
        self.close()
```

MySQLConnection 类的完整代码如下（mysql_conn.py）：

```python
import pymysql

class MySQLConnection(object):
    """
    MySQL 连接
    """
    def __init__(self, host=None, user=None, password=None, database=None, port=3306):
        self.host = host
        self.user = user
        self.password = password
        self.database = database
        self.port = port
        self.db = self.get_db()
        self.conn = self.conn
```

```python
def get_db(self):
    """
    打开数据库连接
    :return:
    """
    db = pymysql.connect(self.host, self.user, self.password, self.database, self.port)
    return db

def conn(self):
    """
    使用 cursor()方法获取操作游标
    :return:
    """
    cursor = self.db.cursor()
    return cursor

def close(self):
    """
    关闭数据库连接
    :return:
    """
    self.db.close()

def query_all(self, query_sql):
    """
    根据查询语句查询所有数据
    :param query_sql:
    :return:
    """
    try:
        cursor = self.conn()
        # 执行 SQL 语句
        cursor.execute(query_sql)
        # 获取所有记录列表
        results = cursor.fetchall()
        return results
    except Exception as ex:
        print("query all Error: {}".format(ex))
    finally:
        self.close()

def query_one(self, query_sql):
    """
    根据查询语句查询一条语句
    :param query_sql:
    :return:
    """
    try:
        cursor = self.conn()
        # 执行 SQL 语句
        cursor.execute(query_sql)
```

```python
            # 获取一条记录
            results = cursor.fetchone()
            return results
        except Exception as ex:
            print("query all Error: {}".format(ex))
        finally:
            self.close()

    def update(self, update_sql):
        """
        根据更新语句更新数据
        :param update_sql:
        :return:
        """
        try:
            cursor = self.conn()
            # 执行 SQL 语句
            cursor.execute(update_sql)
            # 提交到数据库执行
            self.db.commit()
        except Exception as ex:
            print("update Error: {}".format(ex))
            # 发生错误时回滚
            self.db.rollback()
        finally:
            self.close()
```

到此为止,数据库连接、数据查询和数据更新的代码封装完成,接下来看如何使用。

数据查询文件 select_exp.py 内容更改如下(select_exp_01.py):

```python
from chapter3.common.mysql_conn import MySQLConnection


def mysql_select():
    """
    数据查找
    :return:
    """
    # 待查询的学号
    s_num = 1002
    # SQL 查询语句
    sql = "SELECT * FROM python_class WHERE number = {}".format(s_num)
    try:
        # 获得数据库连接对象及游标
        conn = MySQLConnection("localhost", "root", "root", "data_school")
        # 执行 SQL 语句,并获取所有记录列表
        results = conn.query_all(sql)
        for row in results:
            num = row[1]
            name = row[2]
            cs_name = row[3]
            # 打印结果
            print(f"学号为{s_num}的详细信息为:"
```

```
                    f"number={num},name={name},class_name={cs_name}")
        except:
            print("Error: unable to fetch data")

if __name__ == "__main__":
    mysql_select()
```

在 select_exp_01.py 文件中不再需要导入 pymysql 库，直接从 mysql_conn 中导入 MySQLConnection 类即可。初始化 MySQLConnection 类即获得数据库连接对象及游标。

通过代码可知，已经不需要编写获取数据库连接、游标的代码了，也不需要编写数据库关闭的代码。

数据插入文件 insert_exp_02.py 内容更改如下（insert_exp_03.py）：

```
from chapter3.common.mysql_conn import MySQLConnection

def get_conn():
    # 获得数据库连接对象及游标
    conn = MySQLConnection("localhost", "root", "root", "data_school")
    return conn

def insert_record():
    """
    插入数据
    :return:
    """
    # SQL 插入语句
    sql = """INSERT INTO python_class(number,name, class_name)
            VALUES (1006, '小王', 'Python 快乐学习班')"""
    try:
        # 执行 SQL 语句
        get_conn().update(sql)
        print('记录插入成功。')
    except Exception as ex:
        print("update Error: {}".format(ex))

if __name__ == "__main__":
    insert_record()
```

数据更改文件 updata_exp_01.py 内容更改如下（updata_exp_02.py）：

```
from chapter3.common.mysql_conn import MySQLConnection

def get_conn():
    # 获得数据库连接对象及游标
    conn = MySQLConnection("localhost", "root", "root", "data_school")
    return conn

def query_mysql(s_num):
    """
    根据条件查找数据
    :param s_num:
    :return:
```

```python
    """
    # SQL 查询语句
    sql = "SELECT * FROM python_class WHERE number={}".format(s_num)
    try:
        # 执行 SQL 语句,并获取所有记录列表
        results = get_conn().query_all(sql)
        for row in results:
            num = row[1]
            name = row[2]
            # 打印结果
            print(f"学号为{s_num}的详细信息为:number={num},name={name}")
    except Exception as ex:
        print("query Error: {}".format(ex))

def update_mysql():
    """
    数据更新
    :return:
    """
    # SQL 更新语句
    sql = "UPDATE python_class SET name = '{}'  WHERE name='{}'".format('小李', '小张')
    try:
        # 执行 SQL 语句
        get_conn().update(sql)
        # 提交到数据库执行
    except Exception as ex:
        print("update Error: {}".format(ex))

if __name__ == "__main__":
    number = 1005
    print("--------------更改之前---------------")
    query_mysql(number)
    update_mysql()
    print("--------------更改之后---------------")
    query_mysql(number)
```

限于篇幅,本书暂时介绍这几种封装方式。

这里介绍的高级封装也只是一种比较简单的封装方式,大家有兴趣可以探索更简洁和抽象的封装方式。如对数据库访问的 IP 地址、用户名、密码、数据库等内容的读取采用文件配置方式配置,封装一个方法去读取配置文件,从而避免数据库连接的敏感信息泄露。

3.6 小结

本章主要讲解了 Python 通过 PyMySQL 操作 MySQL 数据库的各种基本操作,都是一些实战性的操作。

在本章学习过程中,学有余力的同学可以对本章的内容再做进一步的封装,如加上数据库连接池的封装、对大表操作的封装、对数据库读取超时的封装等。

数据库的操作在数据处理中是非常关键的一环,如果对数据库的操作不熟悉,在数

据处理中就会走很多弯路。需要与数据打交道的工作者掌握好数据库处理技术是非常有必要的。

3.7 实战演练

1. 完成 PyMySQL 环境安装。
2. 通过 PyMySQL 连接第 2 章中创建的数据库，并打印其中一张表的记录。
3. 使用 PyMySQL 对 python_class 表中记录进行插入、删除、修改操作。
4. 结合前面所学，查找网络资源，编写 Python 代码，使用 PyMySQL 对表数据进行排序、分组等操作。
5. 参照 3.4 节的示例，编写代码实现多表操作，并打印出多表操作的结果。
6. 根据 3.5 节的高级封装，将实战 5 的代码进行封装。若根据示例代码封装，再思考是否有其他操作可以进行封装。

第 4 章　SQLAlchemy 的安装和操作

通过 PyMySQL 操作 MySQL 数据库使用的是直接编写 SQL 语句的方式。这种操作方式对于不熟悉 SQL 语句的开发人员是一个比较大的挑战。本章将学习一种更适合程序开发者通过代码操作数据库的框架——SQLAlchemy。

学校为加强同学们对进入 IT 大讲堂听讲座的体验，推出各班级可以在网上预先选座位的服务，但是要求各班以排为单位选择。工作人员根据大家选择的排号，做好各班所选排的标识和秩序维护。各班人员即使不熟悉会场，也可以先选择座位，而工作人员也不需要花大量精力去安排各个班的座位了。

SQLAlchemy 的操作与自行选择座位类似，通过 SQLAlchemy 操作数据库，操作人员可以不熟悉数据库，只需操作映射到数据库的对象即可，具体的数据库操作会通过数据库对象映射到数据库。

4.1　SQLAlchemy 简介

一些程序员因为惧怕 SQL 而在开发的时候小心翼翼地写着 SQL 语句，心中总是少不了恐慌，万一不小心 SQL 语句出错，搞坏了数据怎么办？或者为了获取一些数据，需要了解各种连接，或是需要编写函数、存储过程等。毫无疑问，不搞懂这些，怎么都觉得别扭，说不定什么时候就往"坑"里跳了。

ORM（Object Relational Mapping，对象关系映射）让畏惧 SQL 的开发者在一定程度上得到了解救，他们可以通过 ORM 绕过 SQL。ORM 就是把数据库的一个个 table（表）映射为编程语言的 class（类）。

对 Python 这种面向对象的程序来说，一切皆对象，但是使用的数据库都是关系型的，为了保证一致的使用习惯，通过 ORM 将编程语言的对象模型和数据库的关系模型建立映射关系。在使用编程语言对数据库进行操作的时候，可以直接使用编程语言的对象模型操作数据库，而不使用 SQL 语句。

ORM 把表映射成类，把行作为实例，把字段作为属性，在执行对象操作时最终会把对象的操作转换为数据库原生语句。

Python 中比较著名的 ORM 框架有很多，SQLAlchemy 就是其中一种，并且是 Python 世界里非常好用的一种。

采用写原生 SQL 的方式在代码中会出现大量的 SQL 语句，会出现一些问题：

① SQL 语句重复利用率不高，越复杂的 SQL 语句条件越多，代码越长，会出现越多相近的 SQL 语句。

② 很多 SQL 语句是在业务逻辑中拼出来的，如果有数据库需要更改，就要去修改这些逻辑，这容易漏掉对某些 SQL 语句的修改。

③ SQL 语句容易忽略 Web 安全问题，给未来造成隐患。

而使用 ORM 有如下优点。

① 易用性：使用 ORM 做数据库的开发可以有效地减少重复 SQL 语句的概率，写出来的模型更加直观、清晰。

② 性能损耗小：ORM 转换成底层数据库操作指令确实会有一些开销，但从实际情况来看，这种性能损耗很少。只要不是对性能有严格的要求，综合考虑开发效率、代码的阅读性，使用 ORM 带来的好处要远大于性能损耗，而且项目越大作用越明显。

③ 设计灵活：可以轻松地写出复杂的查询。

④ 可移植性：SQLAlchemy 封装了底层的数据库实现，支持多个关系型数据库引擎，包括流行的 MySQL、PostgreSQL 和 SQLite 等，可以轻松地切换数据库。

SQLAlchemy 分为两部分：ORM 的对象映射和核心的 SQL expression。

ORM 的对象映射是纯粹的 ORM。SQL expression 是 DBAPI 的封装，也提供了很多方法，通过 SQL 表达式来实现。

SQLAlchemy 可以支持 3 种操作方式。

① SQL expression：通过 SQLAlchemy 的方法写 SQL 表达式，间接使用 SQL。

② raw SQL：与 PyMySQL 一样，直接书写 SQL 表达式。

③ ORM：避开直接书写 SQL 表达式。

SQL expression 和 raw SQL 是 ORM 的基础，即使不使用 ORM，两者可以很好地完成工作，并且代码的可读性更好。纯粹把 SQLAlchemy 当成 DB API 使用也可以。

SQLAlchemy 内建数据库连接池，解决了连接操作的烦琐处理，并提供强大的 log 功能。一些复杂的查询语句依靠单纯的 ORM 比较难实现。

SQLAlchemy 代表了用户使用 Python 定义类来与数据库中的表相关联的一种方式，类的实例则对应数据表中的一行数据。

SQLAlchemy 包括了一套将对象中的变化同步到数据库表中的系统，称为工作单元（unit of work），同时提供了使用类查询来实现查询数据库和查询表之间关系的功能。

SQLAlchemy ORM 与 SQLAlchemy 表达式语言（SQLAlchemy Expression Language）是不同的，前者是在后者的基础上构建的，也就是说，是基于后者实现的（将后者封装，类似于实现一套 API）。SQL 表达式语言（SQL Expression Language）代表了关系型数据库最原始的一种架构，而 SQLAlchemy ORM 代表了一种更高级也更抽象的实现，也是 SQL 表达式语言的应用。

一个成功的应用可能只使用 ORM 的构造。在更复杂的情况下，可能 ORM 和 SQL 表达式语言互相配合使用也是必不可少的。

4.2 SQLAlchemy 的安装和连接

4.2.1 安装 SQLAlchemy

在使用 SQLAlchemy 前要先给 Python 安装 MySQL 驱动，由于 MySQL 不支持 Python 3，因此需要使用 PyMySQL 与 SQLAlchemy 交互。

PyMySQL 通过 pip 进行安装。在 Windows 下，使用 pip 安装包时要记得用管理员身份运行 CMD，否则有些操作是无法进行的。在 Linux 和 Mac 下，SQLAlchemy 的安装更简单，此处不详细讲解，有需要的可以自行查找相关资料。

PyMySQL 的安装语句如下：

```
pip install pymysql
```

再安装 SQLAlchemy，语句如下：

```
pip install sqlalchemy
```

安装 SQLAlchemy 后，可以通过如下代码查看版本（view_version.py）：

```
import sqlalchemy

# 查看 SQLAlchemy 版本
print(f'the version of SQLAlchemy:{sqlalchemy.__version__}')
```

输出结果如下：

```
the version of SQLAlchemy:1.2.10
```

至此，SQLAlchemy 安装完成。

4.2.2 使用 SQLAlchemy 连接 MySQL 数据库

使用 SQLAlchemy 连接 MySQL 数据库时，需要通过 import 导入必要的包，示例如下：

```
from sqlalchemy import create_engine
```

其中，from 指的是从 SQLAlchemy 中插入必需的模板。导入包后，需要创建一个连接引擎，示例如下：

```
engine=create_engine("mysql+pymysql://root:root@localhost:3306/data_school", echo=True)
```

连接引擎的表示形式如下：

```
create_engine("数据库类型+数据库驱动://数据库用户名:数据库密码@IP 地址:端口号/数据库", 其他参数)
```

create_engine()方法进行数据库连接，返回一个 db 对象。

方法中的参数解释如下。

① 数据库类型+数据库驱动：连接引擎的标准写法，指定数据库类型和使用的数据库连接的驱动。通过 SQLAlchemy 连接数据库必须通过相应的数据库驱动来连接，如示例中的"mysql+pymysql"是指使用 PyMySQL 驱动连接 MySQL 数据库。

② 数据库用户名：数据库连接的用户名。

③ 数据库密码：数据库的用户密码，若没有设置密码，则密码不需要填写，但密码前面的":"是必需的。

④ IP 地址：数据库主机的地址。若为本地，数据库地址可以写本地 IP 地址或 localhost，非本地则填写对应 IP 地址。

⑤ 端口号：数据库连接的默认状态下端口号通常为 3306。端口号有更改的则需要填写更改后的端口号。

⑥ 数据库：连接的数据库名，如示例中的 data_school。

现在更流行的连接引擎写法如下：

```
engine = create_engine('mysql+pymysql://root:root@localhost:3306/data_school?charset=utf8', echo=True)
```

语句格式如下:

```
create_engine("数据库类型+数据库驱动://数据库用户名:数据库密码@IP 地址:端口号/数据库? 编码", 其他参数)
```

① 编码：数据库的编码方式，一般选择 UTF-8，对中文支持比较好，不易出现乱码。

② echo 参数：用来设置 SQLAlchemy 日志，通过 Python 标准库 logging 模块实现，设置为 True 时，可以看见所有的操作记录。建议将 echo 设置为 false，减少日志的输出。

create_engine()的返回值是 Engine 的一个实例，代表了操作数据库的核心接口，处理数据库和数据库的 API。

创建连接引擎的完整示例如下（engine_create.py）：

```
from sqlalchemy import create_engine

# 创建连接引擎
engine = create_engine('mysql+pymysql://root:root@localhost:3306/data_school?charset=utf8', echo=True)
```

初次调用 create_engine()时并不会真正连接数据库，只有在真正执行一条命令时才会尝试建立连接，目的是节省资源。很多地方都会使用这种方式，如 Python 中的 lazy property 称为 Lazy Connecting（懒惰连接）。

4.2.3 映射声明

当使用 ORM 时，其配置过程主要分为两部分：一是描述要处理的数据库表的信息，二是将 Python 类映射到这些表上。它们在 SQLAlchemy 中一起完成，被称为 Declarative。

使用 Declarative 参与 ORM 映射的类需要被定义为一个指定基类的子类，这个基类应当含有 ORM 映射中相关的类和表的信息。这样的基类称为 declarative base class。

在应用中，一般只需要一个这样的基类。这个基类可以通过 declarative_base 来创建。

映射声明创建代码示例如下（mapping_declarative.py）：

```
from sqlalchemy.ext.declarative import declarative_base

# 声明映射
Base = declarative_base()
```

对应该示例中的 Base 对象经常被称为创建对象的基类。

4.3 SQLAlchemy 常用数据类型

使用 SQLAlchemy 时，声明映射后，基本上就可以使用了，但在使用前先要创建对应的类，而创建类需要类对象，类对象要有属性，属性要有数据类型。

SQLAlchemy 的常用数据类型有如下 12 种。

① Integer：整型，映射到数据库中是 int 类型。

② Float：浮点类型，映射到数据库中是 float 类型，占 32 位。

③ Double：双精度浮点类型，映射到数据库中是 double 类型，占 64 位。

④ String：可变字符类型，映射到数据库中是 varchar 类型。

⑤ Boolean：布尔类型，映射到数据库中是 tinyint 类型。

⑥ Decimal：定点类型，专门为了解决浮点类型精度丢失的问题而设定。存储与货币相关的字段时建议使用该类型。Decimal 类型使用时需要传递两个参数，第一个参数标记该字段能存储多少位数字，第二个参数表示小数点后有多少个小数位。

⑦ Enum：枚举类型，指定某个字段只能是枚举中指定的几个值，不能为其他值。

⑧ Date：日期类型，只能存储年月日，映射到数据库中是 date 类型。在 Python 代码中，可以使用 datetime.date 来指定 Date。

⑨ DateTime：时间类型，存储年月日时分秒毫秒等，映射到数据库中是 datetime 类型。在 Python 代码中，可以使用 datetime.datetime 来指定 DateTime。

⑩ Time：时间类型，只能存储时分秒，映射到数据库中是 time 类型。在 Python 代码中，可以使用 datetime.time 来指定 Time。

⑪ Text：长字符串，可以存储 6 万多个字符，映射到数据库中是 text 类型。如果超出了这个范围，可以使用 Longtext 类型。

⑫ Longtext：长文本类型，映射到数据库中是 longtext 类型。

这 12 种数据类型是 SQLAlchemy 目前版本中的常用数据类型。若在使用中有看到超出这 12 种数据类型的其他类型，需要到 SQLAlchemy 的官网查找是否有新类型的添加，以免引起不必要的错误。

4.4 创建类

前面已经创建了一个基类，并且介绍了常用数据类型，现在可以基于这个基类来创建自定义类了。下面以建立一个课程类为例做进一步的讲解。

从 Base 派生一个名为 Course 的类，其中可以定义将要映射到数据库的表的属性（主要是表的名称、列的类型和名称等）。示例如下（create_table_01.py）：

```python
from sqlalchemy import Column, Integer, String
from sqlalchemy.ext.declarative import declarative_base

# 声明映射
Base = declarative_base()

# 定义 Course 对象，课程表对象
class Course(Base):
    # 表的名称
    __tablename__ = 'course'
    id = Column(Integer, primary_key=True)
    course_name = Column(String(20), default=None, nullable=False, comment='课程名称')
    teacher_name = Column(String(20), default=None, nullable=False, comment='任课老师')
    class_times = Column(Integer, default=0, nullable=False, comment='课时')
```

由代码段可知，Course 类继承了 Base 类。

通过 Declarative 生成的类至少应该包含一个名为 tablename 的属性来给出目标表的名称，以及至少一个 Column 来给出表的主键（primary key）。

Column 类的列选项中必须有常用数据类型声明，以声明这个 Column 是什么类型。除了声明类型，Column 中还有很多列选项可供选择，如表 4-1 所示。

表 4-1 Column 常用列选项

选项名	说明
primary_key	是否为主键，如果为 true，则代表表的主键
unique	是否唯一，如果为 true，则代表这列不允许出现重复的值
index	如果为 true，则为这列创建索引，提高查询效率
nullable	如果为 true，则允许有空值；如果为 false，则不允许有空值
default	为这列定义默认值
name	该属性在数据库中的字段映射
autoincrement	是否自动增长
onupdate	更新时执行的函数
comment	字段描述

SQLAlchemy 不会对类名和表名之间的关联做任何假设，也不会自动设计数据类型和约束的转换。一般地，用户可以自己创建一个模板来建立这些自动转换，这样可以减少很多重复劳动。

比如，可以定义一个函数，该函数返回一个可以用来表示对象的可打印字符串。示例如下（create_table_02.py）：

```python
from sqlalchemy import Column, Integer, String
from sqlalchemy.ext.declarative import declarative_base

# 声明映射
Base = declarative_base()

# 定义 Course 对象，课程表对象
class Course(Base):
    # 表的名字
    __tablename__ = 'course'
    id = Column(Integer, primary_key=True)
    course_name = Column(String(20), default=None, nullable=False, comment='课程名称')
    teacher_name = Column(String(20), default=None, nullable=False, comment='任课老师')
    class_times = Column(Integer, default=0, nullable=False, comment='课时')

    # 定义__repr__函数，返回一个可以用来表示对象的可打印字符串
    def __repr__(self):
        c_name = self.course_name
        t_name = self.teacher_name
        c_times = self.class_times
        return f"Course:(course_name={c_name}, teacher_name={t_name}, class_times={c_times})"
```

由类与对象的学习中知道，当打印一个类对象时，一般只会返回一个地址，不会返回更具体的内容。若能返回一些可查看的信息，就需要在类中自定义一些函数，通过函数返回更直观的信息，如以上示例中的 __repr__ 函数。

类声明完成后，Declarative 会将所有的 Column 成员替换为特殊的 Python 访问器（accessors），称为 Descriptors，这个过程称为 Instrumentation，从而映射成能够读写数据库的表和列。

4.5 创建模式

通过 Declarative 构建好 Course 类后，关于表的信息也已经创建好了，称为 table metadata，描述这些信息的类为 Table。在 Python 交互模式下，可以通过 __table__ 类变量来查看表信息。示例如下（table_metadata.txt）：

```
>>> import sqlalchemy
>>> from sqlalchemy import Column, Integer, String
>>> from sqlalchemy.ext.declarative import declarative_base
>>> Base = declarative_base()
>>> class Course(Base):
    __tablename__ = 'course'
    id = Column(Integer, primary_key=True)
    course_name = Column(String(20), default=None, nullable=False, comment='课程名称')
    teacher_name = Column(String(20), default=None, nullable=False, comment='任课老师')
    class_times = Column(Integer, default=0, nullable=False, comment='课时')

>>> Course.__table__
Table('course', MetaData(bind=None), Column('id', Integer(), table=<course>,
primary_key=True, nullable=False), Column('course_name', String(length=20), table=<course>,
nullable=False), Column('teacher_name', String(length=20), table=<course>, nullable=False),
Column('class_times', Integer(), table=<course>, nullable=False, default=ColumnDefault(0)),
schema=None)
```

注意：这些类的描述信息在 Python 交互模式下才可见。

当完成类声明时，Declarative 用一个 Python 的 metaclass 为这个类进行了加工。在这个阶段，它依据给出的设置创建了 Table 对象，然后构造一个 Mapper 对象与之关联。用户不需与这些幕后的对象直接打交道。

Table 对象是一个更大家庭——称为 metadata 的一部分。当使用 Declarative 时，这个对象也可以在基类的.metadata 属性中看到。

metadata 是与数据库打交道的一个接口。对于 MySQL 数据库而言，此时还没有一个名为 course 的表，需要使用 metadata 发出 CREATE TABLE 的命令。

下面使用 metadata.create_all()指令，将前面得到的 Engine 作为参数传入。如果前面设置了 echo 为 True，则可以从控制台看到该过程中的 SQL 指令。执行 create_all()指令时会先检查 course 表的存在性，如果不存在，则执行表的创建工作。即在 create_table_02.py 文件代码的最后加上如下语句，执行后即可创建表：

```
Base.metadata.create_all(engine)
```

完整代码如下（create_table_03.py）：

```
from sqlalchemy import Column, Integer, String
from sqlalchemy.ext.declarative import declarative_base
from sqlalchemy import create_engine

# 创建连接引擎
engine = create_engine('mysql+pymysql://root:root@localhost:3306/data_school?charset=utf8', echo=True)

# 声明映射
Base = declarative_base()
```

```python
# 定义 Course 对象，课程表对象
class Course(Base):
    # 表的名字
    __tablename__ = 'course'
    id = Column(Integer, primary_key=True)
    course_name = Column(String(20), default=None, nullable=False, comment='课程名称')
    teacher_name = Column(String(20), default=None, nullable=False, comment='任课老师')
    class_times = Column(Integer, default=0, nullable=False, comment='课时')

    # 定义__repr__函数，返回一个可以用来表示对象的可打印字符串
    def __repr__(self):
        c_name = self.course_name
        t_name = self.teacher_name
        c_times = self.class_times
        return f"Course:(course_name={c_name}, teacher_name={t_name}, class_times={c_times})"

Base.metadata.create_all(engine)
```

执行代码后，因为已设置 echo=True，可以看到如下输出结果：

```
2018-11-23 21:05:24,030 INFO sqlalchemy.engine.base.Engine SHOW VARIABLES LIKE 'sql_mode'
2018-11-23 21:05:24,030 INFO sqlalchemy.engine.base.Engine {}
2018-11-23 21:05:24,174 INFO sqlalchemy.engine.base.Engine SELECT DATABASE()
E:\python\python37\lib\site-packages\pymysql\cursors.py:170: Warning:
(1366, "Incorrect string value: '\\xD6\\xD0\\xB9\\xFA\\xB1\\xEA...' for column
'VARIABLE_VALUE' at row 518")
2018-11-23 21:05:24,174 INFO sqlalchemy.engine.base.Engine {} result = self._query(query)
2018-11-23 21:05:24,175 INFO sqlalchemy.engine.base.Engine SHOW collation
                    WHERE 'Charset' = 'utf8mb4' AND 'Collation' = 'utf8mb4_bin'
2018-11-23 21:05:24,175 INFO sqlalchemy.engine.base.Engine {}
2018-11-23 21:05:24,208 INFO sqlalchemy.engine.base.Engine
                    SELECT CAST('test plain returns' AS CHAR(60)) AS anon_1
2018-11-23 21:05:24,208 INFO sqlalchemy.engine.base.Engine {}
2018-11-23 21:05:24,220 INFO sqlalchemy.engine.base.Engine
                    SELECT CAST('test unicode returns' AS CHAR(60)) AS anon_1
2018-11-23 21:05:24,220 INFO sqlalchemy.engine.base.Engine {}
2018-11-23 21:05:24,221 INFO sqlalchemy.engine.base.Engine SELECT CAST
        ('test collated returns' AS CHAR CHARACTER SET utf8mb4) COLLATE utf8mb4_bin AS anon_1
2018-11-23 21:05:24,221 INFO sqlalchemy.engine.base.Engine {}
2018-11-23 21:05:24,223 INFO sqlalchemy.engine.base.Engine DESCRIBE 'course'
2018-11-23 21:05:24,223 INFO sqlalchemy.engine.base.Engine {}
2018-11-23 21:05:24,233 INFO sqlalchemy.engine.base.Engine ROLLBACK
2018-11-23 21:05:24,234 INFO sqlalchemy.engine.base.Engine
CREATE TABLE course (
    id INTEGER NOT NULL AUTO_INCREMENT,
    course_name VARCHAR(20) NOT NULL COMMENT '课程名称',
    teacher_name VARCHAR(20) NOT NULL COMMENT '任课老师',
    class_times INTEGER NOT NULL COMMENT '课时',
    PRIMARY KEY (id)
)

2019-11-23 21:05:24,234 INFO sqlalchemy.engine.base.Engine {}
2019-11-23 21:05:24,778 INFO sqlalchemy.engine.base.Engine COMMIT
```

由输出可知，执行过程会把类转化为原生的表创建语句。

以 create_table_02.py 文件中的代码形式即可通过类创建一个表。执行完示例代码后，查看数据库即可看到生成的表，如图 4-1 所示。

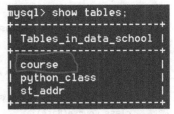

图 4-1 查看表生成结果

4.6 创建映射类的实例

创建好类对象后，就可以通过类对象创建类实例。

对 Course 类创建实例，先尝试不写 __repr__ 函数，看看返回的类对象是怎样的。示例如下（create_table_04.py）：

```python
from sqlalchemy import Column, Integer, String
from sqlalchemy.ext.declarative import declarative_base
from sqlalchemy import create_engine

# 创建连接引擎
engine = create_engine('mysql+pymysql://root:root@localhost:3306/data_school?charset=utf8', echo=True)

# 声明映射
Base = declarative_base()

# 定义 Course 对象，课程表对象
class Course(Base):
    # 表的名字
    __tablename__ = 'course'
    id = Column(Integer, primary_key=True)
    course_name = Column(String(20), default=None, nullable=False, comment='课程名称')
    teacher_name = Column(String(20), default=None, nullable=False, comment='任课老师')
    class_times = Column(Integer, default=0, nullable=False, comment='课时')

# 创建 Course 类实例
course_obj = Course(course_name='Python', teacher_name='Teacher liu', class_times=32)
print(f'Course 对象==>{course_obj}')
print(f'课程名称: {course_obj.course_name}')
print(f'任课老师: {course_obj.teacher_name}')
print(f'课时: {course_obj.class_times}')
```

执行以上代码，得到如下输出结果：

```
Course 对象==><__main__.Course object at 0x0000000003435BE0>
课程名称：Python
任课老师：Teacher Liu
课时：32
```

由结果可知，返回的 Course 对象是一个地址，通过类对象获取具体的属性，打印的属性值是正常的。

在以上代码的基础上加上 __repr__ 函数，更改为如下形式（create_table_05.py）：

```python
from sqlalchemy import Column, Integer, String
from sqlalchemy.ext.declarative import declarative_base
from sqlalchemy import create_engine
```

```python
# 创建连接引擎
engine = create_engine('mysql+pymysql://root:root@localhost:3306/data_school?charset=utf8', echo=True)

# 声明映射
Base = declarative_base()

# 定义 Course 对象，课程表对象
class Course(Base):
    # 表的名字
    __tablename__ = 'course'
    id = Column(Integer, primary_key=True)
    course_name = Column(String(20), default=None, nullable=False, comment='课程名称')
    teacher_name = Column(String(20), default=None, nullable=False, comment='任课老师')
    class_times = Column(Integer, default=0, nullable=False, comment='课时')

    # 定义__repr__函数，返回一个可以用来表示对象的可打印字符串
    def __repr__(self):
        c_name = self.course_name
        t_name = self.teacher_name
        c_times = self.class_times
        return f"Course:(course_name={c_name}, teacher_name={t_name}, class_times={c_times})"

# 创建 Course 类实例
course_obj = Course(course_name='Python', teacher_name='Teacher Liu', class_times=32)
print(f'Course 对象==>{course_obj}')
print(f'课程名称: {course_obj.course_name}')
print(f'任课老师: {course_obj.teacher_name}')
print(f'课时: {course_obj.class_times}')
```

执行以上代码，得到输出结果如下：

```
Course 对象==>Course:(course_name=Python, teacher_name=Teacher Liu, class_times=32)
课程名称: Python
任课老师: Teacher Liu
课时: 32
```

由结果可知，增加了__repr__函数后，打印 Course 类对象时得到了想要的输出结果。同样，类对象的属性值打印也是正常的。

4.7 创建会话

要真正应用类对象操作表，还需要一个 Session 对象。Session 是一个非常重要的概念，ORM 对数据库的入口即 Session。

当构建应用时，在 create_engine 的同一级别下定义一个 Session 类，作为生成新的 Session 的工厂（Factory）类。Session 工厂类的创建方式如下（session_create_01.py）：

```python
from sqlalchemy import create_engine
from sqlalchemy.orm import sessionmaker

# 创建连接引擎
engine = create_engine('mysql+pymysql://root:root@localhost:3306/data_school?charset=utf8', echo=True)
# 创建 Session 类
Session = sessionmaker(bind=engine)
```

```
# 构造新的 Session
session = Session()
```

以上代码方式即可创建 Session 工厂。

若要在创建 engine 前定义 Sesssion，则 bind 可以先不设置，创建 Session 工厂的代码可以更改如下（session_create_02.py）：

```
from sqlalchemy import create_engine
from sqlalchemy.orm import sessionmaker

# 创建 Session
Session = sessionmaker()
# 创建连接引擎
engine = create_engine('mysql+pymysql://root:root@localhost:3306/data_school?charset=utf8', echo=True)
Session.configure(bind=engine)
# 构造新的 Session
session = Session()
```

由更改后的代码可以看到，通过后续定义好的 engine 对象，再通过 configure()将 engine 对象连接到 Session。

有了这个自定义的工厂类后，就可以拿来构造新的 Session 了。

经过上面的操作，Session 已经与 MySQL 数据库的 Engine 关联了。不过目前还没有打开任何到数据库的连接，当一个 Session 被首次使用时，它会从 Engine 所维护的连接池中取出一个连接来操作数据库。这个连接在应用有所更改或者关闭 Session 时会被释放。

SQLAlchemy 的 Session 是用于管理数据库操作的一个像容器一样的工厂对象。Session 工厂对象中提供 query()、add()、add_all()、commit()、delete()、flush()、rollback()、close()等方法，各方法的具体使用会在第 5 章讲解，大概作用介绍如下：

- ❖ query()方法用于数据查询。
- ❖ add()和 add_all()方法用于数据的插入。
- ❖ commit()方法用于事务的提交。
- ❖ delete()方法用于从 Session 中移除某个对象。
- ❖ flush()方法的作用是在事务管理内与数据库发生交互，对应的实例状态被反映到数据库，如自增 ID 被填充上值。
- ❖ rollback()方法的作用是回滚变更。
- ❖ close()方法用于关闭 Session。

通过 SQLAlchemy 的 Session 可实现事务嵌套，由 begin_nested()方法做 savepoint 后，即可实现嵌套。SQLAlchemy 中的事务嵌套有两种情况：一种是在 Session 中管理的事务，本身有层次性；另一种是在 Session 和原始的 connection 之间，是一种层次关系，Session 和 connection 两个概念之中的事务同样具有这样的层次。

通过 SQLAlchemy 的 Session 还可以实现二段式提交。

二段式提交（Two-Phase）：为解决分布式环境下多点事务控制的一套协议。一般事务是 begin 后 commit 结束，而二段式提交的流程上 begin 后是 prepare transaction 'transaction_id'，这时相关事务数据已经持久化了。简单来说，就是事务先保存，再做提交。

SQLAlchemy 对该机制的实现是在构建会话类时加入 twophase 参数，示例如下：

```
Session = sessionmaker(twophase=True)
```

设置 twophase=True 后，会话类就可以根据一些策略，绑定的多个 Engine 可以是多个数据库连接。示例如下（two_phase_exp_01.py）：

```python
from sqlalchemy import create_engine
from sqlalchemy.orm import sessionmaker

# 创建 Session
Session = sessionmaker(twophase=True)
# 创建连接引擎
engine_1 = create_engine('mysql+pymysql://root:root@localhost:3306/test1?charset=utf8', echo=True)
engine_2 = create_engine('mysql+pymysql://root:root@localhost:3306/test2?charset=utf8', echo=True)
Session.configure(bind=engine_1)
Session.configure(binds={Object_1: engine_1, Object_2: engine_2})
# 构造新的 Session
session = Session()
```

示例中的 Object_1、Object_2 指的是不同的类对象，都继承自声明映射对象，具体由自己指定即可。

二段式提交的概念有些不好理解，此处作为一个知识点稍作介绍，有兴趣的读者可以查阅相关资料，做更深入的了解。

4.8 小结

本章主要讲解了 SQLAlchemy 的基本概念，更多的实战内容将在第 5 章进行介绍。

SQLAlchemy 的主要功能是帮助开发者以面向对象的方式操作数据库，在一定程度上，可以帮助不熟悉数据库的开发者较好地操作数据库。

4.9 实战演练

1. 完成 SQLAlchemy 的安装。
2. 通过 SQLAlchemy 建立与 MySQL 数据的连接。
3. 使用 SQLAlchemy 在 MySQL 数据库中创建一张表。

第 5 章 SQLAlchemy 操作 MySQL

本章将通过一些操作示例介绍使用 SQLAlchemy 操作数据库的各种方式。

对于去 IT 大讲堂听讲，开通网上选座后，各班的参会人员也可以自由确定了，有人员的变更或座位的变更，大家都可以在网上点击对应座位号进行更改，由系统自动生成变更结果，不再需要现场工作人员来调配了。

Python 通过 SQLAlchemy 操作 MySQL 类似学生自己更改座位，通过 SQLAlchemy 操作数据库映射对象，而不是具体操作数据库，由 SQLAlchemy 数据库映射对象映射到数据库完成相关操作。

5.1 SQLAlchemy 对 MySQL 数据库的基本操作

如 PyMySQL 可以对 MySQL 数据库进行增、删、改、查操作一样，SQLAlchemy 也可以对 MySQL 数据库进行相应的基本操作。

5.1.1 添加对象

前面已经创建了 Session 工厂，并可以通过工厂创建新的 Session 对象。有了 Session 对象，就可以将类实例对象存入数据库。为了将 Course 对象存入数据库，需要调用 Session 的 add()函数。示例如下（insert_exp_01.py）：

```python
from sqlalchemy.ext.declarative import declarative_base
from sqlalchemy.orm import sessionmaker
from sqlalchemy import create_engine
from sqlalchemy import Column, Integer, String

# 声明映射
Base = declarative_base()
# 创建 Session
Session = sessionmaker()
# 创建连接引擎
engine = create_engine('mysql+pymysql://root:root@localhost:3306/data_school?charset=utf8', echo=False)
Session.configure(bind=engine)

# 构造新的 Session
session = Session()

# 定义 Course 对象，课程表对象
class Course(Base):
    # 表的名字
    __tablename__ = 'course'
    id = Column(Integer, primary_key=True)
    course_name = Column(String(20), default=None, nullable=False, comment= '课程名称')
```

```
    teacher_name = Column(String(20), default=None, nullable=False, comment='任课老师')
    class_times = Column(Integer, default=0, nullable=False, comment='课时')

    # 定义__repr__函数,返回一个可以用来表示对象的可打印字符串
    def __repr__(self):
        c_name = self.course_name
        t_name = self.teacher_name
        c_times = self.class_times
        return f"Course:(course_name={c_name}, teacher_name={t_name}, class_times={c_times})"

# 创建 Course 类实例
course_obj = Course(course_name='Python', teacher_name='Teacher Liu', class_times=32)
# 添加对象
session.add(course_obj)
```

执行以上代码,通过 MySQL 指令控制台查看 course 表,结果如图 5-1 所示。

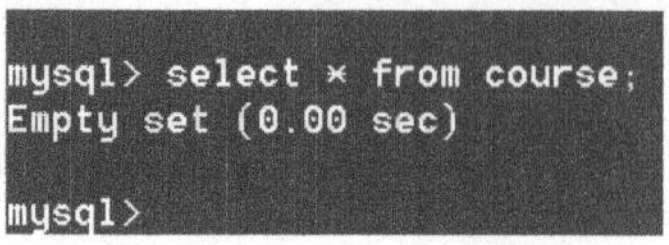

```
mysql> select * from course;
Empty set (0.00 sec)

mysql>
```

图 5-1 插入结果

由查询结果可知,数据库中的 course 表中并没有插入记录。在代码中已经执行了 add 操作,为何在数据库中看不到对应的记录呢?

在 Python 中调用 Session 的 add()方法添加 Course 实例时,调用 add()方法后,所添加的 Course 实例的状态为 pending,就是目前实际上还没有执行 SQL 操作,在数据库中还没有产生与这个 Course 实例对应的行,此时数据是保存在内存中的。若让添加的 Course 实例进入数据库,还需要调用 Session 的 commit()方法提交事务。

将 insert_exp_01.py 中代码片段更改如下 (insert_exp_02.py):

```
from sqlalchemy.ext.declarative import declarative_base
from sqlalchemy.orm import sessionmaker
from sqlalchemy import create_engine
from sqlalchemy import Column, Integer, String

# 声明映射
Base = declarative_base()
# 创建 Session
Session = sessionmaker()
# 创建连接引擎
engine = create_engine('mysql+pymysql://root:root@localhost:3306/data_school?charset=utf8', echo=False)
Session.configure(bind=engine)
# 构造新的 Session
session = Session()

# 定义 Course 对象,课程表对象
class Course(Base):
    # 表的名字
    __tablename__ = 'course'
    id = Column(Integer, primary_key=True)
    course_name = Column(String(20), default=None, nullable=False, comment='课程名称')
```

```
    teacher_name = Column(String(20), default=None, nullable=False, comment='任课老师')
    class_times = Column(Integer, default=0, nullable=False, comment='课时')

    # 定义__repr__函数，返回一个可以用来表示对象的可打印字符串
    def __repr__(self):
        c_name = self.course_name
        t_name = self.teacher_name
        c_times = self.class_times
        return f"Course:(course_name={c_name}, teacher_name={t_name}, class_times={c_times})"

# 创建 Course 类实例
course_obj = Course(course_name='Python', teacher_name='Teacher Liu', class_times=32)
# 添加对象
session.add(course_obj)
# 事务提交
session.commit()
```

由代码片段可以看到，在 insert_exp_01.py 代码文件的最后加上一行事务提交的代码即可。执行代码，通过 MySQL 指令控制台查看 course 表，得到如图 5-2 所示的结果。Course 实例对应的行已经成功插入 MySQL 数据库了。

```
mysql> select * from course;
+----+-------------+--------------+-------------+
| id | course_name | teacher_name | class_times |
+----+-------------+--------------+-------------+
|  1 | Python      | Teacher liu  |          32 |
+----+-------------+--------------+-------------+
1 row in set (0.00 sec)

mysql>
```

图 5-2　插入结果

执行结果已经得到了预期的结果，但是对比 insert_exp_01.py 和 insert_exp_02.py 中的代码可以发现，在这两个 PY 文件中存在大量相同的代码段，并且这些代码都放在一个 PY 文件中，使得代码看上去颇为臃肿。那么，这些代码是否可以抽取出来呢？先抽取 Course 类，为便于后续的操作与使用，在 chapter5 文件夹下创建一个名为 common 的文件夹，在 common 文件夹下创建一个 course.py 文件。文件的代码如下（course.py）：

```
from sqlalchemy import Column, Integer, String
from sqlalchemy.ext.declarative import declarative_base

# 声明映射
Base = declarative_base()

# 定义 Course 对象，课程表对象
class Course(Base):
    # 表的名字
    __tablename__ = 'course'
    id = Column(Integer, primary_key=True)
    course_name = Column(String(20), default=None, nullable=False, comment= '课程名称')
    teacher_name = Column(String(20), default=None, nullable=False, comment='任课老师')
    class_times = Column(Integer, default=0, nullable=False, comment='课时')

    # 定义__repr__函数，返回一个可以用来表示对象的可打印字符串
    def __repr__(self):
```

```
        c_name = self.course_name
        t_name = self.teacher_name
        c_times = self.class_times
        return f"Course:(course_name={c_name}, teacher_name={t_name}, class_times={c_times})"
```

该示例将 Couser 类封装进了一个 PY 文件,将 insert_exp_02.py 更改为调用封装的 Course 类,代码更改如下（insert_exp_03.py）：

```
from sqlalchemy import create_engine
from sqlalchemy.orm import sessionmaker
# 从 course.py 文件中导入 Course 类
from chapter5.common.course import Course

# 创建 Session
Session = sessionmaker()
# 创建连接引擎
engine = create_engine('mysql+pymysql://root:root@localhost:3306/data_school?charset=utf8', echo=True)
Session.configure(bind=engine)
# 构造新的 Session
session = Session()

# 创建 Course 类实例
course_obj = Course(course_name='Python', teacher_name='Teacher liu', class_times=32)
# 添加对象
session.add(course_obj)
# 事务提交
session.commit()
```

由更改后的代码可以看到,更改的 insert_exp_03.py 文件中的代码量已经比前面的 insert_exp_02.py 文件中少了很多,看起来也没有那么臃肿了,不过还是有重复编写的代码,Session 创建的代码还在重复编写。

接下来继续做代码抽取和封装。根据前面的示例代码,抽取一个创建 Session 的公共代码,与 course.py 一样,在 common 文件夹下创建一个名为 session_create.py 的 PY 文件,文件中的代码示例如下（session_create.py）：

```
from sqlalchemy import create_engine
from sqlalchemy.orm import sessionmaker

def get_session():
    """
    创建 session 并返回
    :return: session 对象
    """
    # 创建 Session
    connect_str = 'mysql+pymysql://root:root@localhost:3306/data_school?charset=utf8'
    Session = sessionmaker()
    # 创建连接引擎
    engine = create_engine(connect_str, echo=True)
    Session.configure(bind=engine)
    # 构造新的 Session

    session = Session()
    return session
```

由示例代码可以看到，代码中提供了一个 get_session()方法，返回一个构造好的 Session 对象，外部程序调用 get_session()方法即可拿到 Session 对象。

现在结合 course.py 和 session_create.py 文件，将 insert_exp_02.py 更改为如下更为精简的结构（insert_exp_04.py）：

```python
# 从 session_create.py 文件中导入 get_session 函数
from chapter5.common.session_create import get_session
# 从 course.py 文件中导入 Course 类
from chapter5.common.course import Course

def table_insert():
    # 取得 session 对象
    session = get_session()
    # 创建 Course 类实例
    course_obj = Course(course_name='Python', teacher_name='Teacher liu', class_times=32)
    # 添加对象
    session.add(course_obj)
    # 事务提交
    session.commit()

if __name__ == "__main__":
    table_insert()
```

由代码片段可以看到，现在的代码结构已经变得非常简单清晰了。

此示例代码也不是最精简或不可再封装的，可以进一步精简和封装，此处不再具体讲解，有兴趣的读者可以继续深入探究。

后续的讲解中，当需要用到 Course 类或 Session 的创建时，会直接使用本节创建好的 course.py 和 session_create.py 文件中的代码片段，而不再重复编写这些代码段。若后续有可以抽取的公共代码段，也会将对应的 PY 文件放到 common 文件夹下，便于文件的统一管理。

在 Session 中添加对象的方法，除了 add()方法，还有 add_all()方法。add()方法一次只能添加一个对象，而 add_all()方法可以添加多个对象。add_all()方法的调用方式如下：

```python
session.add_all([object1,object2,object3....])
```

其中，object1、object2、object3 都是一个 Base 的派生类。示例如下（mult_insert_exp.py）：

```python
# 从 session_create.py 文件中导入 get_session 函数
from chapter5.common.session_create import get_session

# 从 course.py 文件中导入 Course 类
from chapter5.common.course import Course

def table_insert():
    # 取得 session 对象
    session = get_session()
    # 创建 Course 类实例
    course_obj_1 = Course(course_name='MySQL', teacher_name='Teacher Wang', class_times=32)
    course_obj_2 = Course(course_name='PyMySQL', teacher_name='Teacher Zhang', class_times=32)
    course_obj_3 = Course(course_name='SQLAlchemy', teacher_name='Teacher Gao', class_times=32)
    # 添加多个对象
    session.add_all([course_obj_1, course_obj_2, course_obj_3])
```

```
    # 事务提交
    session.commit()

if __name__ == "__main__":
    table_insert()
```

执行代码，在日志输出控制台可以看到包含如下输出信息：

```
2018-11-27 19:49:38,313 INFO sqlalchemy.engine.base.Engine {}
2018-11-27 19:49:38,313 INFO sqlalchemy.engine.base.Engine BEGIN (implicit)
2018-11-27 19:49:38,313 INFO sqlalchemy.engine.base.Engine INSERT INTO course (course_name,
 teacher_name, class_times) VALUES (%(course_name)s, %(teacher_name)s, %(class_times)s)
2018-11-27 19:49:38,313 INFO sqlalchemy.engine.base.Engine {'course_name': 'MySQL',
 'teacher_name': 'Teacher Wang', 'class_times': 32}
2018-11-27 19:49:38,313 INFO sqlalchemy.engine.base.Engine INSERT INTO course (course_name,
 teacher_name, class_times) VALUES (%(course_name)s, %(teacher_name)s, %(class_times)s)
2018-11-27 19:49:38,313 INFO sqlalchemy.engine.base.Engine {'course_name': 'PyMySQL',
 'teacher_name': 'Teacher Zhang', 'class_times': 32}
2018-11-27 19:49:38,313 INFO sqlalchemy.engine.base.Engine INSERT INTO course (course_name,
 teacher_name, class_times) VALUES (%(course_name)s, %(teacher_name)s, %(class_times)s)
2018-11-27 19:49:38,313 INFO sqlalchemy.engine.base.Engine {'course_name': 'SQLAlchemy',
 'teacher_name': 'Teacher Gao', 'class_times': 32}
2018-11-27 19:49:38,313 INFO sqlalchemy.engine.base.Engine COMMIT
```

由输出信息可以看到，在 SQLAlchemy 的操作中，执行 Session 的 add_all()方法时，实际会将多个对象的操作转化为一条条 INSERT 语句，逐条往数据库中插入数据。

程序执行后，通过 MySQL 指令控制台查看执行结果，可以看到如图 5-3 所示的结果。

```
mysql> select * from course;
+----+-------------+---------------+-------------+
| id | course_name | teacher_name  | class_times |
+----+-------------+---------------+-------------+
|  1 | Python      | Teacher liu   |          32 |
|  2 | MySQL       | Teacher Wang  |          32 |
|  3 | PyMySQL     | Teacher Zhang |          32 |
|  4 | SQLAlchemy  | Teacher Gao   |          32 |
+----+-------------+---------------+-------------+
4 rows in set (0.00 sec)

mysql>
```

图 5-3　add_all 操作结果

由结果可以看到，对应的几个 Couser 对象都成功插入 course 表中了。

在实际项目应用中，能以面向对象的方式创建对象是非常不错的，可以避免在源代码中加入 SQL 代码。

5.1.2 查询对象

在 Session 中添加对象的方法时，每次执行完插入操作，都需要到 MySQL 指令控制台查看数据，这样操作很不方便。Session 中提供了 query()方法实现数据查询。

Session 的 query()方法会返回一个 Query 对象。query()方法可以接受多种参数类型，可以是类或者类的属性字段。

如要通过 Session 的 query()方法从 course 表中查找出所有的记录，查找所有记录的方式还需要在 query()方法后加上 all()方法，代码如下（query_exp_1.py）：

```python
# 从 session_create.py 文件中导入 get_session 函数
from chapter5.common.session_create import get_session
# 从 course.py 文件中导入 Course 类
from chapter5.common.course import Course


def table_query():
    """
    表数据查询
    :return:
    """
    # 取得 session 对象
    session = get_session()
    # 数据查询
    query_result = session.query(Course).all()
    for item in query_result:
        print(f'查询结果为==>{item}')


if __name__ == "__main__":
    table_query()
```

执行该代码段，在日志输出控制台可以看到包含如下输出：

```
2018-11-27 20:35:45,033 INFO sqlalchemy.engine.base.Engine BEGIN (implicit)
2018-11-27 20:35:45,033 INFO sqlalchemy.engine.base.Engine SELECT course. id AS course_id,
 course.course_name AS course_course_name, course.teacher_name AS course_teacher_name,
 course.class_times AS course_class_times
FROM course
2018-11-27 20:35:45,033 INFO sqlalchemy.engine.base.Engine {}
查询结果为==>Course:(course_name=Python, teacher_name=Teacher Liu, class_ times=32)
查询结果为==>Course:(course_name=MySQL, teacher_name=Teacher Wang, class_ times=32)
查询结果为==>Course:(course_name=PyMySQL, teacher_name=Teacher Zhang, class_times=32)
查询结果为==>Course:(course_name=SQLAlchemy, teacher_name=Teacher Gao, class_times=32)
```

由结果可知，通过 Session 的 query()方法和 all()方法，可以拿到所有的表记录，并且编译器会将查询语句转换为对应的 SQL 语句。

由打印的日志结果同时可以看到，在 Course 类中添加__repr__()方法对于打印指定形式的结果的便捷性。

这里可以拿到全部记录，若只查找 course_name 为 Python 的记录，查找方式怎样呢？

只查找 course_name 为 Python 的记录，在 SQL 语句中需要使用 WHERE 条件语句实现，在 SQLAlchemy 中，可以使用 query()方法中的 filter()方法实现与 WHERE 等同的效果。具体使用看如下示例（query_exp_2.py）：

```python
# 从 session_create.py 文件中导入 get_session 函数
from chapter5.common.session_create import get_session
# 从 course.py 文件中导入 Course 类
from chapter5.common.course import Course


def table_query():
    """
    表数据查询
    :return:
```

```python
    """
    # 取得session对象
    session = get_session()
    # 数据查询
    query_result = session.query(Course).filter(Course.course_name == 'Python')

    for item in query_result:
        print(f'查询结果为==>{item}')

if __name__ == "__main__":
    table_query()
```

执行代码文件，在日志输出控制台可以看到包含如下输出：

```
2018-11-27 20:49:16,419 INFO sqlalchemy.engine.base.Engine BEGIN (implicit)
2018-11-27 20:49:16,419 INFO sqlalchemy.engine.base.Engine SELECT course.id AS course_id,
 course.course_name AS course_course_name, course.teacher_name AS course_teacher_name,
 course.class_times AS course_class_times
FROM course
WHERE course.course_name = %(course_name_1)s
2018-11-27 20:49:16,419 INFO sqlalchemy.engine.base.Engine {'course_name_1': 'Python'}
查询结果为==>Course:(course_name=Python, teacher_name=Teacher Liu, class_times=32)
```

由结果可知，查询得到了想要的结果，并且编译器将加了 filter() 方法的查询语句转换为了 SQL 的 WHERE 条件查询语句。

以面向对象的方式进行对象查询，在实际项目应用中，对代码的维护带来很大的方便，也可以避免很多查询风险，如 SQL 注入。若可能，在项目中多使用面向对象的方式编写查询语句。

5.1.3 更新对象

在 SQLAlchemy 中，通过 Session 对象做数据的更新操作是先得到数据表对应的类对象，若要更改某个字段值，找到字段映射到类对象对应的属性字段，对属性字段赋值为需要更改的值，再执行 Session 的 add() 方法并执行 commit() 后，即实现更新。

若需要将数据表 course 中，对 course_name 为 Python 的记录，将 teacher_name 的值更改为 "Teacher Li"，则代码如下（update_exp_1.py）：

```python
# 从session_create.py文件中导入get_session函数
from chapter5.common.session_create import get_session
# 从course.py文件中导入Course类
from chapter5.common.course import Course

def table_update():
    """
    数据查询并修改
    :return:
    """

    # 取得session对象
    session = get_session()
    # 数据查询
    query_result = session.query(Course).filter(Course.course_name == 'Python')
```

```python
        for item in query_result:
            print(f'查询结果为==>{item}')
            print(f'item对象的类型为:{type(item)}')
            print(f'更改前 teacher_name={item.teacher_name}')
            # 将 Teacher liu 更改为 Teacher Li
            item.teacher_name = 'Teacher Li'
            session.add(item)
            session.commit()

        query_result = session.query(Course).filter(Course.course_name == 'Python')
        for item in query_result:
            print(f'更改后 teacher_name={item.teacher_name}')

if __name__ == "__main__":
    table_update()
```

执行代码文件，在日志控制台中可以看到包含如下输出：

```
2018-11-27 21:47:23,066 INFO sqlalchemy.engine.base.Engine BEGIN (implicit)
2018-11-27 21:47:23,066 INFO sqlalchemy.engine.base.Engine SELECT course. id AS course_id,
 course.course_name AS course_course_name, course.teacher_name AS course_teacher_name,
 course.class_times AS course_class_times
FROM course
WHERE course.course_name = %(course_name_1)s
2018-11-27 21:47:23,066 INFO sqlalchemy.engine.base.Engine {'course_name_1': 'Python'}
查询结果为==>Course:(course_name=Python, teacher_name=Teacher Liu, class_times=32)
item对象的类型为:<class 'chapter5.common.course.Course'>
更改前: teacher_name=Teacher Liu
2018-11-27 21:47:23,076 INFO sqlalchemy.engine.base.Engine UPDATE course
 SET teacher_name=%(teacher_name)s WHERE course.id = %(course_id)s
2018-11-27 21:47:23,076 INFO sqlalchemy.engine.base.Engine {'teacher_name': 'Teacher Li',
 'course_id': 1}
2018-11-27 21:47:23,076 INFO sqlalchemy.engine.base.Engine COMMIT
2018-11-27 21:47:23,248 INFO sqlalchemy.engine.base.Engine BEGIN (implicit)
2018-11-27 21:47:23,248 INFO sqlalchemy.engine.base.Engine SELECT course. id AS course_id,
 course.course_name AS course_course_name, course.teacher_name AS course_teacher_name,
 course.class_times AS course_class_times

FROM course
WHERE course.course_name = %(course_name_1)s
2018-11-27 21:47:23,248 INFO sqlalchemy.engine.base.Engine {'course_ name_1': 'Python'}
更改后: teacher_name=Teacher Li
```

在输出日志信息中可以看到如下关键信息：

```
查询结果为==>Course:(course_name=Python, teacher_name=Teacher Liu, class_ times=32)
item对象的类型为:<class 'chapter5.common.course.Course'>
更改前: teacher_name=Teacher Liu

更改后: teacher_name=Teacher Li
```

这里，输出查询结果是为了更清晰地查看所查询到的结果的形式是怎样的，输出 item 对象的类型是为了清楚地了解 item 是什么结构类型。

打印更改前和更改后 teacher_name 的值，能得知更改操作是否发生，也免于再到

MySQL 指令控制台去进行查验。

从输出的日志可知，编译器将 Session 发起的 add()方法编译成了 UPDATE 语句。

对于 update_exp_1.py 文件中的示例代码，可更改如下（update_exp_2.py）：

```python
# 从 session_create.py 文件中导入 get_session 函数
from chapter5.common.session_create import get_session
# 从 course.py 文件中导入 Course 类
from chapter5.common.course import Course

def table_update():
    """
    数据查询并修改
    :return:
    """
    # 取得 session 对象
    session = get_session()
    # 数据查询
    course_obj = session.query(Course).filter(Course.course_name == 'Python').first()
    print(course_obj)
    print(type(course_obj))
    print(f'更改前: teacher_name={course_obj.teacher_name}')
    # 将 Teacher Li 更改为 Teacher Liu
    course_obj.teacher_name = 'Teacher Liu'
    session.add(course_obj)
    session.commit()

    course_obj = session.query(Course).filter(Course.course_name == 'Python').first()
    print(course_obj)
    print(f'更改后: teacher_name={course_obj.teacher_name}')

if __name__ == "__main__":
    table_update()
```

在代码中，在 filter()后有使用到 first()方法，意为取得查询结果中的第一个结果对象，如示例中拿到的就是一个 Course 对象。得到 Course 对象后就可以直接进行更改操作，如代码中所示。

直接运行代码，即可看到更改的结果，此处不再展示更改结果。

对于示例代码，有兴趣的读者可以进一步简化或封装。

5.1.4 删除对象

在 SQLAlchemy 中，通过 Session 对象做对象删除操作比较简单。先查询要删除的数据，得到的数据映射为类对象，再调用 Session 的 delete()方法删除类对象，执行后即完成删除操作。

如将 course 表中 course_name 为"Python"的记录删除，代码如下（delete_exp_1.py）：

```python
# 从 session_create.py 文件中导入 get_session 函数
from chapter5.common.session_create import get_session
# 从 course.py 文件中导入 Course 类
from chapter5.common.course import Course
```

```python
def table_delete():
    """
    表数据查询
    :return:
    """
    # 取得 session 对象
    session = get_session()
    # 数据查询
    course_obj = session.query(Course).filter(Course.course_name == 'Python').first()
    # print(course_obj)
    print(f'删除前==>{course_obj}')
    # 数据删除
    session.delete(course_obj)
    session.commit()

    course_obj = session.query(Course).filter(Course.course_name == 'Python').first()
    print(f'删除后==>{course_obj}')

if __name__ == "__main__":
    table_delete()
```

执行代码文件，在日志输出控制台可以看到包含如下输出日志：

```
2018-11-27 22:22:49,608 INFO sqlalchemy.engine.base.Engine BEGIN (implicit)
2018-11-27 22:22:49,618 INFO sqlalchemy.engine.base.Engine SELECT course. id AS course_id,
 course.course_name AS course_course_name, course.teacher_name AS course_teacher_name,
 course.class_times AS course_class_times
FROM course
WHERE course.course_name = %(course_name_1)s
 LIMIT %(param_1)s
2018-11-27 22:22:49,618 INFO sqlalchemy.engine.base.Engine {'course_name_1': 'Python', 'param_1': 1}
删除前==>Course:(course_name=Python, teacher_name=Teacher Liu, class_ times=32)
2018-11-27 22:22:49,618 INFO sqlalchemy.engine.base.Engine DELETE FROM course WHERE course.id = %(id)s
2018-11-27 22:22:49,618 INFO sqlalchemy.engine.base.Engine {'id': 1}
2018-11-27 22:22:49,649 INFO sqlalchemy.engine.base.Engine COMMIT
2018-11-27 22:22:49,806 INFO sqlalchemy.engine.base.Engine BEGIN (implicit)
2018-11-27 22:22:49,806 INFO sqlalchemy.engine.base.Engine SELECT course. id AS course_id,
 course.course_name AS course_course_name, course.teacher_name AS course_teacher_name,
 course.class_times AS course_class_times
FROM course
WHERE course.course_name = %(course_name_1)s
 LIMIT %(param_1)s
2018-11-27 22:22:49,806 INFO sqlalchemy.engine.base.Engine {'course_name_1': 'Python', 'param_1': 1}
删除后==>None
```

由输出日志可知，对应记录已经成功删除。同时，操作过程中，编译器会将 delete()方法对应的语句转换为 SQL 的 DELETE 语句。

5.2 SQLAlchemy 的常用 filter 操作符

在 5.1.2 节中，代码示例中用到了 all()、first()等方法，本节将进一步介绍 filter()方法中常用的操作符。

5.2.1　equals 操作符

在 filter()方法中，equals 操作符在 query_exp_2.py 中已有使用，即如下代码行：

```
query_result = session.query(Course).filter(Course.course_name == 'Python')
```

代码行中的"=="操作符就是 equals 操作符，意为等于操作，操作结果是 True 或 False。若操作符两边的值相等，则结果是 True，否则结果是 False。

在实际项目应用中，equals 操作是很常用的一个操作，类似 SQL 语句中的等值操作。

5.2.2　not equals 操作符

not equals 操作符即不等于操作符，结果也是 True 或 False。

not equals 的示例如下（not_equals_exp.py）：

```python
# 从 session_create.py 文件中导入 get_session 函数
from chapter5.common.session_create import get_session
# 从 course.py 文件中导入 Course 类
from chapter5.common.course import Course

def table_query():
    """
    表数据查询
    :return:
    """
    # 取得 session 对象
    session = get_session()
    # 数据查询，查找 course 表中 course_name 不等于 MySQL 的记录
    query_result = session.query(Course).filter(Course.course_name != 'MySQL')
    for item in query_result:
        print(f'查询结果为==>{item}')

if __name__ == "__main__":
    table_query()
```

该示例代码用于从 course 表中查找 course_name 不等于'MySQL'的记录。执行该示例代码，可以看到如下输出结果：

```
2018-11-28 22:30:38,235 INFO sqlalchemy.engine.base.Engine SELECT course. id AS course_id,
course.course_name AS course_course_name, course.teacher_name AS
course_teacher_name, course.class_times AS course_class_times
FROM course
WHERE course.course_name != %(course_name_1)s

2018-11-28 22:30:38,235 INFO sqlalchemy.engine.base.Engine {'course_name_ 1': 'MySQL'}
查询结果为==>Course:(course_name=PyMySQL, teacher_name=Teacher Zhang, class_times=32)
查询结果为==>Course:(course_name=SQLAlchemy, teacher_name=Teacher Gao, class_times=32)
```

由输出结果可以看到，没有返回 course_name 等于'MySQL'的记录，同时对象操作的语句转换成了 SQL 查询语句。

后续的操作示例代码中，由于与 not_equals_exp.py 文件中的代码相差不大，后面的几个示例只展示有差异的代码，具体代码可以根据给定的 PY 文件去查看，也会适当展示执行结果。

5.2.3 like 操作符

在 filter() 方法中，like 操作与 SQL 中的 LIKE 操作类似，作用也是模糊匹配，有全模糊匹配、左模糊匹配和右模糊匹配三种，示例代码分别如下（like_exp.py）。

全模糊匹配：

```
# 全模糊匹配,,teacher_name 中包含 Wang 这个字符串记录都查询出来
query_result = session.query(Course).filter(Course.teacher_name.like ('%Wang%'))
```

执行代码文件，可以看到如下输出：

```
2018-11-28 22:59:36,964 INFO sqlalchemy.engine.base.Engine SELECT course. id AS course_id,
 course.course_name AS course_course_name, course.teacher_name AS course_teacher_name,
 course.class_times AS course_class_times
 FROM course
 WHERE course.teacher_name LIKE %(teacher_name_1)s
2018-11-28 22:59:36,964 INFO sqlalchemy.engine.base.Engine {'teacher_name_ 1': '%Wang%'}
查询结果为==>Course:(course_name=MySQL, teacher_name=Teacher Wang, class_ times=32)
```

右模糊匹配：

```
# 右模糊匹配，teacher_name 中包含以字符串 Wang 结尾的记录都查询出来
query_result = session.query(Course).filter(Course.teacher_name.like ('%Wang'))
```

执行代码文件，可以看到如下输出：

```
2018-11-28 23:00:08,680 INFO sqlalchemy.engine.base.Engine SELECT course. id AS course_id,
 course.course_name AS course_course_name, course.teacher_name AS
 course_teacher_name, course.class_times AS course_class_times
 FROM course
 WHERE course.teacher_name LIKE %(teacher_name_1)s
2018-11-28 23:00:08,680 INFO sqlalchemy.engine.base.Engine {'teacher_name_ 1': 'Wang%'}
```

由输出可以看到，没有得到匹配的结果。

左模糊匹配：

```
# 左模糊匹配，teacher_name 中包含以 Wang 这个字符串结尾的记录都查询出来
query_result = session.query(Course).filter(Course.teacher_name.like ('%Wang'))
```

执行代码文件，可以看到如下输出：

```
2018-11-28 23:01:09,499 INFO sqlalchemy.engine.base.Engine SELECT course. id AS course_id,
 course.course_name AS course_course_name, course.teacher_name AS course_teacher_name,
 course.class_times AS course_class_times
 FROM course
 WHERE course.teacher_name LIKE %(teacher_name_1)s
2018-11-28 23:01:09,500 INFO sqlalchemy.engine.base.Engine {'teacher_name_ 1': '%Wang'}
查询结果为==>Course:(course_name=MySQL, teacher_name=Teacher Wang, class_ times=32)
```

这里仅仅作为一个结果演示，更多可以自行操作，以观察输出结果的差异。

5.2.4 and 操作符

在 filter()方法中,and 操作符表示"并且"的意思,其前后的条件都满足时,结果是 True,否则结果是 False。

and 操作符有如下 3 种表示方式(and_exp.py)。

方法一:使用 and_()(注意 and 后面还有一个下划线"_")。

```
# 方法一: 使用 and_()
query_result = session.query(Course).\
    filter(and_(Course.teacher_name == 'Teacher Wang', Course.course_name == 'MySQL'))
```

方法二:在 filter()中设置多个表达式。

```
# 方法二: 在 filter()中设置多个表达式
query_result = session.query(Course).\
    filter(Course.teacher_name == 'Teacher Wang', Course.course_name == 'MySQL')
```

方法三:使用多个 filter()。

```
# 方法三: 使用多个 filter()
query_result = session.query(Course).filter(Course.teacher_name == 'Teacher Wang').\
    filter(Course.course_name == 'MySQL')
```

5.2.5 or 操作符

在 filter()方法中,or 操作符表示"或者"的意思,其前后的条件都不满足时,结果是 False,否则结果是 True。

or 操作符的示例如下(or_exp.py):

```
# 使用 or_()
query_result = session.query(Course).\
    filter(or_(Course.teacher_name == 'Teacher Wang', Course.course_name == 'SQLAlchemy'))
```

5.2.6 is null 操作符

在 filter()方法中,is null 操作符用于判断查询的对象是否是空的,并且在 Python 中,None 字符和数据库表记录中的 null 对应。is null 操作符的示例如下(is_null_exp.py):

```
# 使用 is null
query_result = session.query(Course).filter(Course.teacher_name is None)
```

5.2.7 is not null 操作符

在 filter()方法中,is not null 的意思与 is null 相反,用于判断一个查询出来的对象不是空的。is not null 操作符的示例如下(is_not_null_exp.py):

```
# 使用 is not null
query_result = session.query(Course).filter(Course.teacher_name is not None)
```

5.2.8 in 操作符

在 filter()方法中，in 操作符的作用类似 or，一般后面为 list 类型的值。不过，不同于 or 操作符需要前后都有条件值，in 中可以只有一个或多个元素。in 操作符的示例如下（in_exp.py）（注意 in 后还有一个下划线"_"）：

```
# in_查询
query_result = session.query(Course).filter(Course.teacher_name.in_(['Teacher Wang', 'Teacher Zhang']))
```

5.2.9 not in 操作符

在 filter()方法中，not in 操作符的作用与 in 相反。在 SQLAlchemy 中，not in 操作符使用"~"表示。not in 操作符的示例如下（not_in_exp.py）：

```
# not in 查询
query_result = session.query(Course).filter(~Course.teacher_name.in_(['Teacher Wang', 'Teacher Zhang']))
```

5.3 SQLAlchemy 的更多操作

本节将介绍更多 Session 操作数据库的方法。

5.3.1 返回列表和单项

5.1 节中的查询对象操作中用到了 all()方法，更新对象操作中用到了 first()方法，对于 all()方法的查询结果，需要使用 for 循环遍历，而 first()方法的查询结果不需要遍历。

那么，all()方法和 first()方法分别返回的是什么类型的结果？

在 SQLAlchemy 中，all()方法返回的是一个列表，列表中的每个元素对应一个类对象或类的属性对象。first()方法返回至多一个结果，而且以单个类对象或属性对象的形式，而不是以只包含一个元素的 Tuple 的形式返回结果。

在 SQLAlchemy 的返回查询结果的方法中还有一个 one()方法。

先通过如下示例了解 one()方法的使用（query_one_exp.py）：

```
# 从 session_create.py 文件中导入 get_session 函数
from chapter5.common.session_create import get_session
# 从 course.py 文件中导入 Course 类
from chapter5.common.course import Course


def table_query():
    """
    表数据查询
    :return:
    """
    # 取得 session 对象
    session = get_session()
    # 返回多于一个查询结果
```

```python
        query_result = session.query(Course).one()
        print(f'查询结果为==>{query_result}')

if __name__ == "__main__":
    table_query()
```

文件中的查询语句会返回多个查询结果，所以执行代码文件后，会得到如下错误信息：

```
sqlalchemy.orm.exc.MultipleResultsFound: Multiple rows were found for one()
```

把查询语句更改如下：

```python
query_result = session.query(Course).filter(Course.course_name == 'Python').one()
```

执行代码，会得到如下错误信息：

```
sqlalchemy.orm.exc.NoResultFound: No row was found for one()
```

接着更改查询语句为如下形式：

```python
query_result = session.query(Course).filter(Course.course_name == 'MySQL').one()
```

执行代码，最终得到如下正确返回结果：

```
查询结果为==>Course:(course_name=MySQL, teacher_name=Teacher Wang, class_times=32)
```

由以上执行结果可知：one()方法返回且仅返回一个查询结果，当结果的数量不足一个或者多于一个时会报错。

5.3.2 嵌入使用 SQL

在 SQLAlchemy 中查询对象时，可以在 query()方法中通过使用 text()使用 SQL 语句。使用 text()方法需要先从 SQLAlchemy 中导入 text。操作示例如下（query_text_exp_1.py）：

```python
# 从 session_create.py 文件中导入 get_session 函数
from chapter5.common.session_create import get_session
# 从 course.py 文件中导入 Course 类
from chapter5.common.course import Course
from sqlalchemy import text

def table_query():
    """
    表数据查询
    :return:
    """

    # 取得 session 对象
    session = get_session()
    # 数据查询
    query_result = session.query(Course).filter(text("course_name='MySQL'")).all()
    for item in query_result:
        print(f'查询结果为==>{item}')

if __name__ == "__main__":
    table_query()
```

执行代码段，打印日志中可以看到包含如下输出信息：

```
2018-11-29 22:34:04,732 INFO sqlalchemy.engine.base.Engine SELECT course.id AS course_id,
  course.course_name AS course_course_name, course.teacher_name AS course_teacher_name,
  course.class_times AS course_class_times
FROM course
WHERE course_name='MySQL'
2018-11-29 22:34:04,732 INFO sqlalchemy.engine.base.Engine {}
查询结果为==>Course:(course_name=MySQL, teacher_name=Teacher Wang, class_times=32)
```

观察输出的打印语句，仔细阅读查询的 SQL 语句，WHERE 条件后面的 course_name 前没有加上"course."字符，此处即使用了原生的 SQL 语句。

除了这种直接将参数写进字符串的方式，还可以通过 params() 方法来传递参数。示例如下（query_text_exp_2.py）：

```python
# 从 session_create.py 文件中导入 get_session 函数
from chapter5.common.session_create import get_session
# 从 course.py 文件中导入 Course 类
from chapter5.common.course import Course
from sqlalchemy import text

def table_query():
    """
    表数据查询
    :return:
    """
    # 取得 Session 对象
    session = get_session()
    # 数据查询
    query_result = session.query(Course).\
        filter(text("teacher_name=:t_name and course_name=:c_name")).\
        params(t_name='Teacher Wang', c_name='MySQL').all()

    for item in query_result:
        print(f'查询结果为==>{item}')

if __name__ == "__main__":
    table_query()
```

执行代码段，打印日志中可以看到包含如下输出信息：

```
2018-11-29 22:42:43,120 INFO sqlalchemy.engine.base.Engine SELECT course.id AS course_id,
  course.course_name AS course_course_name, course.teacher_name AS course_teacher_name,
  course.class_times AS course_class_times
FROM course
WHERE teacher_name=%(t_name)s and course_name=%(c_name)s
2018-11-29 22:42:43,120 INFO sqlalchemy.engine.base.Engine {'t_name':'Teacher Wang', 'c_name': 'MySQL'}
查询结果为==>Course:(course_name=MySQL, teacher_name=Teacher Wang, class_times=32)
```

观察输出的打印语句，仔细阅读查询的 SQL 语句，WHERE 条件后面的 teacher_name 和 course_name 前都没有加上"course."字符，也是使用了原生的 SQL 语句。

text() 方法中还可以直接使用完整的 SQL 语句。示例如下（query_text_exp_3.py）：

```python
# 从 session_create.py 文件中导入 get_session 函数
from chapter5.common.session_create import get_session
# 从 course.py 文件中导入 Course 类
from chapter5.common.course import Course
from sqlalchemy import text

def table_query():
    """
    表数据查询
    :return:
    """
    # 取得 Session 对象
    session = get_session()
    # 数据查询
    query_result = session.query(Course).\
        from_statement(text("SELECT * FROM course WHERE course_name= 'MySQL'"))
    for item in query_result:
        print(f'查询结果为==>{item}')

if __name__ == "__main__":
    table_query()
```

执行代码段,打印日志中可以看到包含如下输出信息:

```
2018-11-29 22:47:37,768 INFO sqlalchemy.engine.base.Engine SELECT * FROM course WHERE course_name='MySQL'
2018-11-29 22:47:37,768 INFO sqlalchemy.engine.base.Engine {}
查询结果为==>Course:(course_name=MySQL, teacher_name=Teacher Wang, class_times=32)
```

观察输出的打印语句,仔细阅读查询打印出的 SQL 语句,可以看到此处输出的是原生的 SQL 语句。

在 text() 方法中可以使用原生的 SQL 语句,但是使用时要注意将表名和列名写正确。

5.3.3 计数

在 SQL 语句中可以使用 count(*) 的形式进行记录数的统计,那么在 SQLAlchemy 中要对记录做统计,又该怎么操作呢?

SQLAlchemy 的 query 定义了一个方便的计数方法 count()。示例如下(query_count_exp_1.py):

```python
# 从 session_create.py 文件中导入 get_session 函数
from chapter5.common.session_create import get_session
# 从 course.py 文件中导入 Course 类
from chapter5.common.course import Course
from sqlalchemy import func

def table_query():
    """
    表数据查询
    :return:
    """
    # 取得 Session 对象
    session = get_session()
```

```
# 数据记录统计
query_result = session.query(Course).count()
print(f'查询结果为==>{query_result}')

if __name__ == "__main__":
    table_query()
```

执行代码，打印日志，可以看到包含如下输出信息：

```
2018-11-29 23:00:48,271 INFO sqlalchemy.engine.base.Engine SELECT count(*) AS count_1
FROM (SELECT course.id AS course_id, course.course_name AS course_course_name,
course.teacher_name AS course_teacher_name, course.class_times AS course_class_times
FROM course) AS anon_1

2018-11-29 23:00:48,271 INFO sqlalchemy.engine.base.Engine {}
查询结果为==>3
```

从打印日志中可知，SQLAlchemy 使用 count() 方法解析出来的 SQL 语句有些复杂，是否有更简单的方式呢？

在 SQLAlchemy 中提供了一个 func 函数，其中的 func.count() 更简便。将 query_count_exp_1.py 文件中的查询语句更改如下：

```
query_result = session.query(func.count('*')).select_from(Course).scalar()
```

更改后执行代码，可以看到解析出的 SQL 语句变成如下简单形式了：

```
2018-11-29 23:08:38,749 INFO sqlalchemy.engine.base.Engine SELECT count(% (count_2)s) AS count_1
FROM course
```

若对 course 表的主键进行计数，则 select_from 也可以省略，查询语句可更改成如下形式：

```
query_result = session.query(func.count(Course.id)).scalar()
```

更改后执行代码，可以看到解析出的 SQL 语句变成如下更简单的形式了：

```
2018-11-29 23:12:21,269 INFO sqlalchemy.engine.base.Engine SELECT count (course.id) AS count_1
FROM course
```

由以上几个示例可以了解到：在 SQLAlchemy 中，总是将被计数的查询打包成一个子查询，然后对这个子查询进行计数。即便是最简单的 SELECT count(*) FROM table 语句也会如此处理。为了更精细地控制计数过程，可以采用 func.count() 函数。

5.4 小结

本章主要讲解的是通过 SQLAlchemy 具体操作 MySQL 数据库的一些操作。

本章所讲解的内容属于实际应用中操作比较多的，也相对比较基础。

本章所讲解的内容属于实践性内容，所以在理论上并没有用更多的篇幅去描述。要更好理解本章讲解内容，动手实践是最好的方式。

本章没有讲解 SQLAlchemy 中建立关系、使用 JOIN 进行查询、子查询等比较高级的操作，这些操作都是基于基本操作实现的。对本章基本操作熟悉了的同学可以去思考并实践这些更高级的操作。

5.5 实战演练

1. 模仿书中示例,实现数据库的增、删、改、查。
2. 从数据库中找到 course_name 为"Python"的记录,然后将 course_name 为"Python"的记录修改为"Python3"。
3. 操作 Course 对象,在一个 query 语句中实现 and 和 or 操作。
4. 操作 Course 对象,实现排序操作。
5. 操作 Course 对象,实现分组操作。

第 6 章 MongoDB 基础

本章将介绍非关系型数据库，即 NoSQL，最具代表意义的是 MongoDB。

数字校园有一座后山，名叫 mongo 山。mongo 山上长满了各种植物，各种植物以非常不规则的方式生长在山上各个地方。没有一种植物以一块一块的方式生长，也没有成排或成列生长的，都是以一种非常自然的个体对象形式存在。同学们上山玩耍时，可以将某棵野草拔起，也可以将某棵树木做一番修剪，还可以将某类植物全部铲除，这些操作都不会对其他个体对象或其他植物类产生影响。

MongoDB 中数据的存放如 mongo 山上的植物，没有排或列的概念，只有集合、文档等概念。MongoDB 中的数据库如一座 mongo 山，集合如 mongo 山上的各种植物，可以有多个，文档如 mongo 山上的个体对象，有各种属性。删除文档犹如拔除某棵野草，修改文档犹如修剪某棵树木，删除集合犹如铲除某类植物等。

6.1 MongoDB 简介

MongoDB 是跨平台的、基于分布式文件存储的开源数据库系统，由 C++语言编写而成。用 MongoDB 创建的数据库具备性能高、可用性强、易于扩展等特点。

MongoDB 将数据存储为一个文档，数据结构由键值（key-value）对组成。MongoDB 文档类似 JSON 对象。字段值可以包含字符串、数字和数组，也可以是另一个文档。文档可以嵌套。

MongoDB 有如下特色：

① MongoDB 是一个面向文档存储的数据库，操作起来比较简单容易。
② 可以在 MongoDB 记录中设置任何属性的索引来实现更快的排序。
③ 可以通过本地或者网络创建数据镜像，这使得 MongoDB 有更强的扩展性。
④ 如果负载增加（需要更多的存储空间和更强的处理能力），它可以分布在计算机网络中的其他节点上，即支持分片。
⑤ MongoDB 支持丰富的查询表达式。查询指令使用 JSON 形式的标记，可轻易查询文档中内嵌的对象及数组。MongoDB 支持对文档执行动态查询，使用的是一种不逊色于 SQL 的基于文档的查询语言。
⑥ MongoDB 使用 update()可以实现替换完成的文档（数据）或者一些指定的数据字段。
⑦ MongoDB 允许在服务端执行脚本，可以用 JavaScript 编写某个函数，直接在服务端执行，也可以把函数的定义存储在服务端，下次直接调用。
⑧ MongoDB 支持各种编程语言，如 Python、C、Java、C++、Go、Scala 等。
⑨ MongoDB 安装简单，也易于扩展。

MongoDB 主要有如下适用场景：

① 大数据领域，特别是对非结构化数据的处理中，使用 MongoDB 这种 NoSQL 来存储数据是很通用的操作。

② 内容管理及交付，其中会使用很多非结构化的数据。

③ 移动及社会化基础设施。

④ 用户数据管理。

⑤ 数据中心。

MongoDB 应用案例如下：

① Craiglist 上使用 MongoDB 存档数十亿条记录。

② FourSquare，基于位置的社交网站，在 Amazon EC2 的服务器上使用 MongoDB 分享数据。

③ Shutterfly，以互联网为基础的社会和个人出版服务，使用 MongoDB 的各种持久性满足数据存储的要求。

④ Intuit 公司，一个为小企业和个人提供服务的软件提供商，为小型企业使用 MongoDB 存储用户数据。

⑤ sourceforge.net，资源查找网站，并创建和发布免费开源软件，使用 MongoDB 做后端存储。

⑥ etsy.com，一个购买和出售手工制作品网站，使用 MongoDB 做数据存储。

⑦ 纽约时报，领先的在线新闻门户网站之一，使用 MongoDB 存储数据。

⑧ CERN，著名的粒子物理研究所，欧洲核子研究中心大型强子对撞机的数据使用 MongoDB 存储。

6.2 MongoDB 的安装

MongoDB 有 Windows 版、Linux 版和 MacOS 版，安装时根据操作系统进行选择。MongoDB 的官网下载地址为：https://www.mongodb.com/download-center/community，页面如图 6-1 所示。

图 6-1 MongoDB 下载官网

官网中有三个下拉框，分别是 Version（版本）、OS（操作系统）、Package（压缩包格式）。选择对应的版本、操作系统、压缩包格式后，单击"Download"按钮，即可开始

MongoDB 安装包的下载。或者直接复制 "Download" 按钮下方的地址来下载 MongoDB 安装包（如图 6-1 箭头所指的地址）。

相对于 Windows 操作系统，Linux 和 MacOS 操作系统下安装 MongoDB 比较简单，本节介绍 Windows 操作系统的 MongoDB 安装。

MongoDB 的安装步骤如下。

（1）如图 6-1 所示，选择 Windows 操作系统的 4.0.4 版，是当前的最新版本，压缩包是 MSI 格式的。下载后，得到名为 mongodb-win32-x86_64-2008plus-ssl-4.0.4-signed.msi 的安装包。

（2）打开安装包，弹出如图 6-2 所示的对话框；单击 "运行" 按钮，弹出如图 6-3 所示的窗口；单击 "Next" 按钮，弹出如图 6-4 所示的窗口。

图 6-2　MongoDB 安装（一）

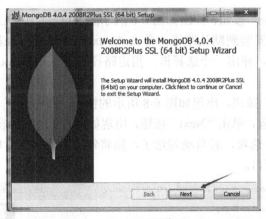

图 6-3　MongoDB 安装（二）

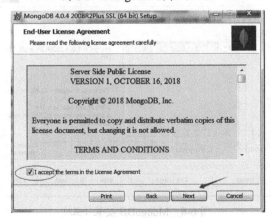

图 6-4　MongoDB 安装（三）

(3) 勾选图 6-4 中的 "I accept the terms in the License Agreement" 复选框，然后单击 "Next" 按钮，弹出如图 6-5 所示的窗口。其中有两种安装方式：Complete 和 Custom。Complete 是默认安装方式，会以默认方式安装到 C 盘，一般不建议以这种方式安装，建议以 Custom 方式安装。

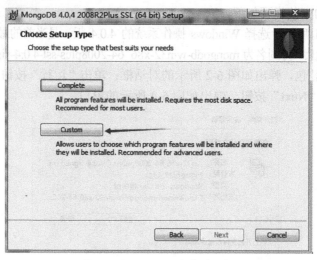

图 6-5　MongoDB 安装（四）

(4) 单击 "Custom" 按钮，弹出如图 6-6 所示的窗口。其中，Location 是指 MongoDB 安装的默认路径，若安装到默认路径，单击 "Next" 按钮即可。若安装到指定路径，则单击 "Browse" 按钮，弹出一个选择框，指定路径（如 E:\mongodb\install\）后，出现如图 6-7 所示的窗口。

(5) 单击 "Next" 按钮，出现如图 6-8 所示的窗口。其中会有一些默认勾选项和默认值，这里使用默认设置，单击 "Next" 按钮，出现如图 6-9 所示的窗口。不要勾选 "Install MongoDB Compass" 选项，若自动勾选了，则将勾选去除。然后单击 "Next" 按钮，出现如图 6-10 所示的窗口。

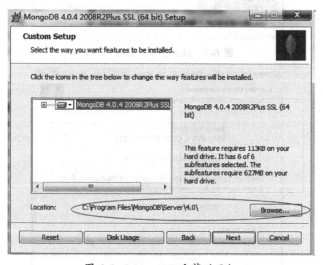

图 6-6　MongoDB 安装（五）

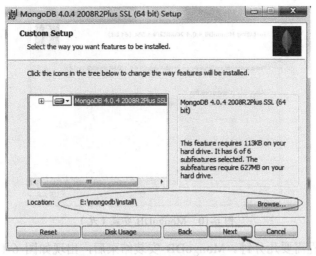

图 6-7 MongoDB 安装（六）

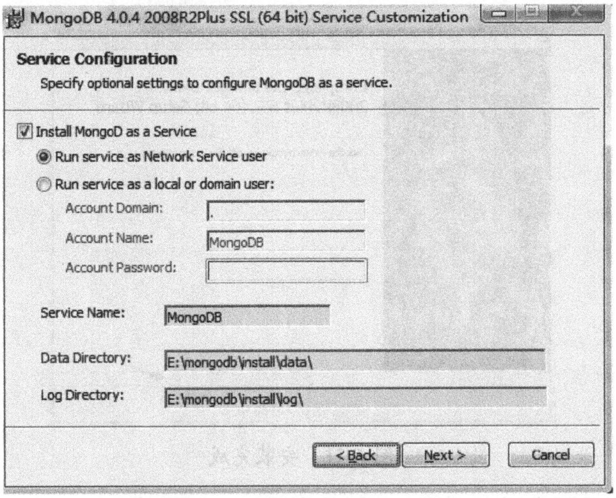

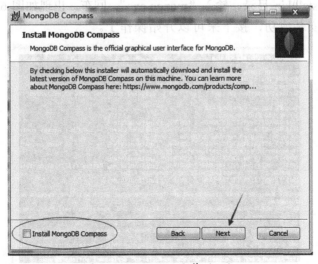

图 6-8 MongoDB 安装（七）

图 6-9 MongoDB 安装（八）

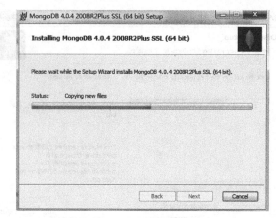

图 6-10 MongoDB 安装（九）

（6）安装过程需要几分钟。MongoDB 安装结束后，出现如图 6-11 所示的窗口，即表明 MongoDB 安装成功。

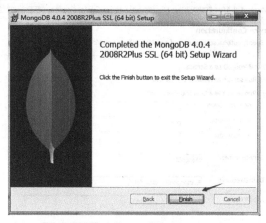

图 6-11 安装完成

在命令提示符界面中进入 MongoDB 安装目录 bin 目录，如 E:\mongodb\install\，则进入 E:\mongodb\install\bin 目录，输入"mongo.exe"，回车，出现如图 6-12 所示内容，即表明 MongoDB 安装成功，接下来可以开始操作了。

图 6-12 命令行界面

在图 6-9 中若勾选了"Install MongoDB Compass",如图 6-13 所示,单击"Next"按钮后,在出现的窗口中单击"Install"按钮,出现如图 6-14 所示的窗口。

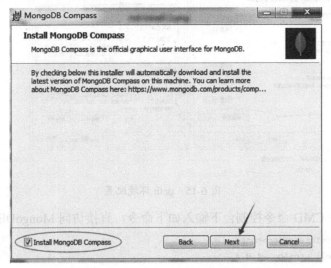

图 6-13　勾选了"Install MongoDB Compass"

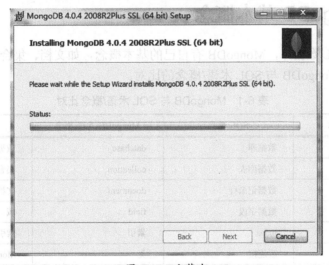

图 6-14　安装中

安装会一直停留在安装中,需要等待非常长的时间,甚至一直结束不了,这时应单击"Cancel"按钮,结束安装。需从头开始安装,并在图 6-9 中不要勾选"Install MongoDB Compass"。

安装好 MongoDB 后,每次要访问 MongoDB 命令控制台,都需要切换到 MongoDB 安装目录的 bin 目录下,操作比较麻烦。

可以通过配置 path 路径来达到更简单的访问方式。操作方式如下:右键单击"我的电脑",在弹出的快捷菜单中选择"属性",在弹出的窗口中选择"高级系统设置→高级→环境变量设置",然后在弹出的对话框中修改"path"变量值,添加 MongoDB 的 bin 目录的全路径目录,如"E:\mongodb\install\bin",如图 6-15 所示。

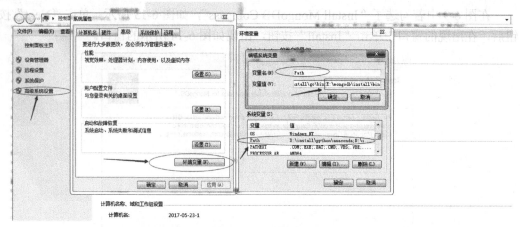

图 6-15　path 环境配置

然后可以在 CMD 命令控制台下输入如下命令，直接访问 MongoDB：

```
C:\Users\Administrator>mongo
MongoDB shell version v4.0.4
```

6.3　MongoDB 基本概念

作为 NoSQL 数据库，MongoDB 有自己的基本概念，如文档、集合等。

表 6-1 是 MongoDB 与 SQL 术语/概念的比对。

表 6-1　MongoDB 与 SQL 术语/概念比对

SQL 术语/概念	SQL 说明	MongoDB 术语/概念	MongoDB 解释
database	数据库	database	数据库
table	数据库表	collection	集合
row	数据记录行	document	文档
column	数据字段	field	域
index	index	索引	索引
table joins	表连接	无	MongoDB 不支持
primary key	主键	primary key	MongoDB 将_id 字段设置为主键

6.3.1　文档

在 MongoDB 中，文档是一组键值（key-value）对，即 BSON。

MongoDB 的文档不需要设置相同的字段，并且相同的字段不需要相同的数据类型，这与关系型数据库有很大的区别，也是 MongoDB 非常突出的特点。

一个简单的文档例子如下：

```
{"class_name":"Python 快乐学习班", "lession":"MongoDB study"}
```

对于文档，需要注意如下几点：

① 文档中的键值对是有序的。
② 文档中的值不仅可以是在双引号中的字符串，还可以是其他数据类型（甚至可以是整个嵌入的文档）。
③ MongoDB 区分类型和大小写。
④ MongoDB 的文档不能有重复的键。
⑤ 文档的键是字符串。除了少数例外情况，键可以使用任意 UTF-8 字符。

文档中键的命名规范如下：
① 键不能含有'\0'（空字符），因为它用来表示键的结尾。
② $有特别的意义，只有在特定环境下才能使用。
③ 以"_"开头的键是保留的（不是严格要求的）。

【知识扩展：BSON】

BSON 是一种计算机数据交换格式，主要被用作 MongoDB 数据库中的数据存储和网络传输格式。它是一种二进制表示形式，用来表示简单数据结构、关联数组（MongoDB 中称为"对象"或"文档"）和 MongoDB 中的各种数据类型。BSON 之名缘于 JSON，含义为 Binary JSON（二进制 JSON）。

6.3.2 集合

集合就是 MongoDB 文档组，类似关系数据库管理系统中的表。

集合存于数据库中，没有固定的结构。这意味着对集合可以插入不同格式和类型的数据，但通常情况下，插入集合的数据都会有一定的关联性。

集合的命名要注意如下几点：
① 集合名不能是空字符串""。
② 集合名不能含有'\0'字符（空字符），因为它表示集合的结尾。
③ 集合名不能以"system."开头，这是为系统集合保留的前缀。
④ 用户创建的集合名字不能含有保留字符。有些驱动程序的确支持在集合名中包含保留字符，这是因为某些系统生成的集合中包含该字符。除非访问这种系统创建的集合，否则千万不要在名字里出现保留字符。

6.3.3 数据库

MongoDB 中可以建立多个数据库。安装好 MongoDB 后，会创建一个默认数据库，数据库名为"db"，存储在 data 目录中。

MongoDB 的单个实例可以容纳多个独立的数据库，每个数据库都有自己的集合和权限，不同的数据库放置在不同的文件中。

在 MongoDB 命令控制台下，执行"DB"命令可以显示当前数据库对象或集合。例如：

```
> DB
test
```

以上示例中输出的 test 即为当前数据库对象。

在 MongoDB 命令控制台下，输入"SHOW dbs"命令可以显示所有数据的列表。

例如：

```
> SHOW dbs
admin   0.000GB
config  0.000GB
local   0.000GB
```

MongoDB 中有些数据库名是保留的，可以直接访问这些有特殊作用的数据库。

① admin：从权限的角度来看，这是 root 数据库。如果将一个用户添加到这个数据库，那么这个用户自动继承所有数据库的权限。一些特定的服务器端命令也只能从这个数据库运行，如列出所有的数据库或者关闭服务器。

② local：这个数据库永远不会被复制，可以用来存储限于本地单台服务器的任意集合。

③ config：当 MongoDB 用于分片设置时，config 数据库在内部使用，用于保存分片的相关信息。

6.3.4 数据类型

MongoDB 的常见数据类型如表 6-2 所示。

表 6-2　MongoDB 的常见数据类型

数据类型	描述
String	字符串，用于存储数据常用的数据类型。在 MongoDB 中，UTF-8 编码的字符串才是合法的
Integer	整型数值，用于存储数值。根据所采用的服务器，可分为 32 位或 64 位
Boolean	布尔值，用于存储布尔值（真/假）
Double	双精度浮点值，用于存储浮点值
Min/Max keys	将一个值与 BSON（二进制的 JSON）元素的最低值和最高值对比
Array	将数组或列表或多个值存储为一个键
Timestamp	时间戳，记录文档修改或添加的具体时间
Object	内嵌文档
Null	创建空值
Symbol	符号，基本等同于字符串类型，但不同的是，它一般用于采用特殊符号类型的语言
Date	日期时间，用 UNIX 时间格式来存储当前日期或时间。可以指定自己的日期时间：创建 Date 对象，传入年月日信息
Object ID	对象 ID，用于创建文档的 ID
Binary Data	二进制数据，用于存储二进制数据
Code	代码类型，用于在文档中存储 JavaScript 代码
Regular expression	正则表达式类型，用于存储正则表达式

1．Object ID

Object ID 类似唯一主键，可以很快地生成和排序，包含 12 字节，含义如下。

① 前 4 字节表示创建 UNIX 时间戳，即格林尼治时间 UTC，比北京时间晚 8 个小时。

② 接下来的 3 字节是机器标识码。

③ 紧接的 2 字节由进程 id 组成，即 PID。

④ 最后 3 字节是由随机数生成的计数器。

具体如图 6-17 所示。

0	1	2	3	4	5	6	7	8	9	10	11
时间戳				机器标识码			PID		计数器		

图 6-17 Object ID 含义

MongoDB 中存储的文档必须有一个 _id 键。其值可以是任何类型的,默认是 Object ID 对象。

由于 Object ID 中保存了创建的时间戳,因此可以不为文档保存时间戳字段。

2. String（字符串）

BSON 字符串都是 UTF-8 编码。

3. Timestamp（时间戳）

BSON 有一个特殊的时间戳类型用于 MongoDB 内部使用,与普通的日期类型不相关。时间戳值是一个 64 位的值,其中前 32 位是一个 time_t 值（与 UNIX 新纪元相差的秒数）,后 32 位是在某秒中操作的一个递增的序数。在单个 mongod 实例中,时间戳的值通常是唯一的。

在复制集中,oplog 有一个 ts 字段,其中的值使用 BSON 时间戳表示操作时间。

BSON 时间戳类型主要用于 MongoDB 内部使用。大多数应用开发中可以使用 BSON 日期类型。

4. Date（日期）

Date 表示当前距离 UNIX 新纪元（1970 年 1 月 1 日）的毫秒数。Date 类型是有符号的,负数表示 1970 年之前的日期。

6.4 MongoDB 基本操作

MongoDB 的基本操作包括:数据库的创建与删除,集合的创建与删除,文档的增、删、改、查等操作。

6.4.1 创建数据库

MongoDB 创建数据库的语法格式如下:

```
USE database_name
```

如果数据库不存在,则创建数据库,否则切换到指定数据库。

先使用如下命令查看当前数据库状态:

```
> SHOW dbs
admin   0.000GB
config  0.000GB
local   0.000GB
> DB
test
```

由输出结果可知，当前没有创建新的数据库。

接下来创建一个名为 mongo_use 的数据库，操作如下：

```
> USE mongo_use
switched to db mongo_use
```

使用 DB 命令查看，操作如下：

```
> DB
mongo_use
```

使用 SHOW dbs 命令查看所有数据库，操作如下：

```
> SHOW dbs
admin    0.000GB
config   0.000GB
local    0.000GB
```

由打印结果可知，刚创建的数据库 mongo_use 并不在数据库的列表中，要显示它，需要向 mongo_use 数据库中插入一些数据。操作如下：

```
> db.mongo_use.insert({"name":"mongodb learning"})
WriteResult({ "nInserted" : 1 })
```

继续查看所有数据库，操作如下：

```
> SHOW dbs
admin       0.000GB
config      0.000GB
local       0.000GB
mongo_use   0.000GB
```

由输出结果可知，此时通过 SHOW dbs 命令可以看到刚才新建的 mongo_use 数据库了。

注意：在 MongoDB 中，集合只有在内容插入后才会创建，也就是说，创建集合（数据表）后要再插入一个文档（记录），集合才会真正创建。

MongoDB 中默认的数据库为 test，如果没有创建新的数据库，集合将存放在 test 数据库中。

6.4.2 删除数据库

MongoDB 删除数据库的语法格式如下：

```
db.dropDatabase()
```

该语句将删除当前数据库。在删除数据库前可以使用 DB 命令查看当前数据库名，若没有指定数据库名，则一般默认数据库是 test。

若需要删除 6.4.1 节中创建的 mongo_use 数据库，操作如下。首先，查看所有数据库，操作如下：

```
> SHOW dbs
admin       0.000GB
config      0.000GB
local       0.000GB
mongo_use   0.000GB
```

接着切换到数据库 mongo_use，操作如下：

```
> USE mongo_use
switched to db mongo_use
```

最后执行数据库删除，操作如下：

```
> db.dropDatabase()
{ "dropped" : "mongo_use", "ok" : 1 }
```

执行删除命令后，用 SHOW dbs 查看所有数据库，操作如下：

```
> SHOW dbs
admin    0.000GB
config   0.000GB
local    0.000GB
```

由打印结果可知，mongo_use 数据库已经不存在了，表明已成功删除。

6.4.3 创建集合

MongoDB 中使用 createCollection()方法来创建集合。其语法格式如下：

```
db.createCollection(name, options)
```

参数说明如下。

name：要创建的集合名称。

options：可选参数，指定有关内存大小及索引的选项，如表 6-3 所示。

表 6-3 options 可选值

参数	类型	描述
capped	布尔	（可选）如为 true，则创建固定集合。固定集合是指有着固定大小的集合，当达到最大值时，它会自动覆盖最早的文档
autoIndexId	布尔	（可选）如为 true，则自动在_id 字段创建索引。默认为 false
size	数值	（可选）为固定集合指定一个最大值（以字节计）。如 capped 为 true，也需指定该字段
max	数值	（可选）指定固定集合中包含文档的最大数量

在插入文档时，MongoDB 首先检查固定集合的 size 字段，然后检查 max 字段。

在 mongo_use 数据库中创建 python_class 集合，操作如下：

```
> USE mongo_use
switched to db mongo_use
> DB
mongo_use
> db.createCollection("python_class")
{ "ok" : 1 }
```

如果查看已有集合，可以使用 SHOW collections 命令，操作如下：

```
> SHOW collections
python_class
```

由打印结果可知，已经创建一个名为 python_class 的集合。

在 MongoDB 中可以使用如下方式创建集合，操作如下：

```
> db.try.insert({"name":"test collection create"})
WriteResult({ "nInserted" : 1 })
```

即不需要先创建集合，当插入一些文档时，MongoDB 会自动创建集合。

查看已有集合情况，操作如下：

```
> SHOW collections
python_class
try
```

由打印结果可知，新增了一个名为 try 的集合。

6.4.4 删除集合

MongoDB 中使用 drop()方法来删除集合，语法格式如下：

```
db.collection.drop()
```

参数说明：collection 为指定需要删除的集合。

返回值：如果成功删除选定集合，则 drop()方法返回 true，否则返回 false。

接下来从 mongo_use 数据库中删除集合 try，操作如下。先通过 SHOW collections 命令查看 mongo_use 数据库中已存在的集合，操作如下：

```
> USE mongo_use
switched to db mongo_use
> SHOW collections
python_class
try
```

再删除集合 try，操作如下：

```
> db.try.drop()
True
```

由返回结果可知，操作返回了一个 True。

通过 SHOW collections 命令再次查看数据库 mongo_use 中的集合：

```
> SHOW collections
python_class
```

由打印结果可知，try 集合已被删除。

6.4.5 插入文档

MongoDB 使用 insert()或 save()方法向集合中插入文档，语法如下：

```
db.collection_name.insert(document)
```

参数说明如下。

collection_name：集合名称。

document：文档。

文档的数据结构与 JSON 基本一样。所有存储在集合中的数据都是 BSON 格式。

向 MongoDB 的 mongo_use 数据库的 python_class 集合中插入文档，操作如下：

```
> USE mongo_use
```

```
switched to db mongo_use
> SHOW collections
python_class
> db.python_class.insert({name:"小萌",class_name:"python 快乐学习班", number:1001})
WriteResult({ "nInserted" : 1 })
```

查看集合 python_class 中已插入文档，操作如下：

```
> db.python_class.find()
{ "_id" : ObjectId("5c1b6e00b9444ea171d76ba8"), "name" : "小萌", "class_name" :"python 快乐学习班", "number" : 1001 }
```

也可以将数据定义为一个变量，操作如下：

```
> document=({name:"小明",class_name:"Python 快乐学习班",number:1002})
```

执行后，显示结果如下：

```
{ "name" : "小明", "class_name" : "Python 快乐学习班", "number" : 1002 }
```

行插入操作：

```
> db.python_class.insert(document)
WriteResult({ "nInserted" : 1 })
```

查看集合 python_class 中已插入文档，操作如下：

```
> db.python_class.find()
{ "_id" : ObjectId("5c1b6e00b9444ea171d76ba8"), "name" : "小萌", "class_ name" :"Python 快乐学习班", "number" : 1001 }
{ "_id" : ObjectId("5c1b6f6bb9444ea171d76ba9"), "name" : "小明", "class_ name" :"Python 快乐学习班", "number" : 1002 }
```

由打印结果可知，document 变量指定的值已经插入文档中。

插入文档也可以使用 db.python_class.save(document)命令。如果不指定_id 字段，则 save()方法类似 insert()方法；如果指定_id 字段，save()方法则会更新该_id 的数据。

6.4.6 更新文档

MongoDB 使用 update()和 save()方法来更新集合中的文档。这两个函数的应用有些区别，下面分别进行介绍。

1. update()方法

update()方法用于更新已存在的文档，语法格式如下：

```
db.collection.update(
    <query>,
    <update>,
    {
        upsert: <boolean>,
        multi: <boolean>,
        writeConcern: <document>
    }
)
```

参数说明如下。

query：更新的查询条件，类似 SQL update 查询语句中 WHERE 后面的查询条件。

update：更新的对象和操作符（如$、$inc）等，类似 SQL 的 UPDATE 查询语句中 SET 后面需要更新字段的值的设置。

upsert：可选，说明如果不存在 update 的记录，是否插入 objNew。若为 true，表示插入；默认是 false，即不插入。

multi：可选，MongoDB 默认是 false，只更新找到的第一条记录，如果其值为 true，就把按条件查找出来的多条记录全部更新。

writeConcern：可选，抛出异常的级别。

若需要将 mongo_use 数据库 python_class 集合中 name 为"小明"的文档更改为"小王"，操作如下：

```
> use mongo_use
switched to db mongo_use
> show collections
python_class
> db.python_class.find()
{ "_id" : ObjectId("5c1b6e00b9444ea171d76ba8"), "name" : "小萌", "class_name" :"Python 快乐学习班", "number" : 1001 }
{ "_id" : ObjectId("5c1b6f6bb9444ea171d76ba9"), "name" : "小明", "class_name" :"Python 快乐学习班", "number" : 1002 }
> db.python_class.update({name:"小明"},{$set:{name:"小王"}})
WriteResult({ "nMatched" : 1, "nUpserted" : 0, "nModified" : 1 })
> db.python_class.find()
{ "_id" : ObjectId("5c1b6e00b9444ea171d76ba8"), "name" : "小萌", "class_name" :"Python 快乐学习班", "number" : 1001 }
{ "_id" : ObjectId("5c1b6f6bb9444ea171d76ba9"), "name" : "小王", "class_name" :"Python 快乐学习班", "number" : 1002 }
```

由打印结果可知，name 为"小明"的文档中的 name 更新为了"小王"。

以上语句只会修改第一条发现的文档，如果修改多条相同的文档，则需设置 multi 参数为 true。如将 class_name 更改为"Python 学习班"，操作如下：

```
> db.python_class.update({class_name:"Python 快乐学习班"},{$set:{class_name:"Python 学习班"}},{multi:true})
WriteResult({ "nMatched" : 2, "nUpserted" : 0, "nModified" : 2 })
> db.python_class.find()
{ "_id" : ObjectId("5c1b6e00b9444ea171d76ba8"), "name" : "小萌", "class_ name" :"Python 学习班", "number" : 1001 }
{ "_id" : ObjectId("5c1b6f6bb9444ea171d76ba9"), "name" : "小王", "class_ name" :"Python 学习班", "number" : 1002 }
```

由打印结果可知，class_name 都由"Python 快乐学习班"更改为了"Python 学习班"。

2. save()方法

save()方法通过传入的文档来替换已有文档，语法格式如下：

```
db.collection.save(
   <document>,
   {
      writeConcern: <document>
   }
)
```

参数说明如下。

document：文档数据。

writeConcern：可选，抛出异常的级别。

如将 mongo_use 数据库 python_class 集合中 _id 为 5c1b6f6bb9444ea171d76b 的文档的 name（即"小王"）更改为"小明"，操作如下：

```
> db.python_class.save({"_id":ObjectId("5c1b6f6bb9444ea171d76ba9"), name: "小明", class_name: "Python 学习班", number:1002})
WriteResult({ "nMatched": 1, "nUpserted": 0, "nModified": 1 })
```

执行操作后，通过 find()命令查看替换后的数据，操作如下：

```
> db.python_class.find()
{ "_id": ObjectId("5c1b6e00b9444ea171d76ba8"), "name": "小萌", "class_name": "Python 学习班", "number": 1001 }
{"_id" : ObjectId("5c1b6f6bb9444ea171d76ba9"), "name": "小明", "class_name": "Python 学习班", "number": 1002 }
```

由打印结果可知，name 为"小王"的文档中的 name 更新为了"小明"。

6.4.7 删除文档

MongoDB 用 remove()函数来移除集合中的数据。

在执行 remove()函数前先执行 find()命令来判断执行的条件是否正确，这是一个比较好的习惯。

remove()方法的语法格式如下：

```
db.collection.remove(
   <query>,
   {
     justOne: <boolean>,
     writeConcern: <document>
   }
)
```

参数说明如下。

query（可选）：删除的文档的条件。

justOne（可选）：如设为 true 或 1，则只删除一个文档，如不设置该参数或使用默认值 false，则删除所有匹配条件的文档。

writeConcern（可选）：抛出异常的级别。

在 mongo_use 数据库的 python_class 集合中查找所有数据，操作如下：

```
> db.python_class.find()
{ "_id" : ObjectId("5c1b95f18ef102160306167a"), "name" : "小萌", "class_ name" :"python 学习班", "number" : 1001 }
{ "_id" : ObjectId("5c1b96148ef102160306167b"), "name" : "小明", "class_ name" :"python 学习班", "number" : 1002 }
```

接下来移除 name 为"小明"的文档，操作如下：

```
> db.python_class.remove({name:"小明"})
WriteResult({ "nRemoved" : 1 })
```

查看集合中的数据：

```
> db.python_class.find()
{ "_id" : ObjectId("5c1b95f18ef102160306167a"), "name" : "小萌", "class_name" :"python 学习班", "number" : 1001 }
```

由打印结果可知，集合中 name 为"小明"的文档已经被删除。

如果想删除 python_class 集合中的所有文档，可以使用以下方式（类似常规 SQL 的 TRUNCATE 命令）：

```
>db.python_class.remove({})
```

在 mongo_use 数据库的 python_class 集合中重复插入一条 name 为"小萌"的文档，操作如下：

```
> db.python_class.insert({name:"小萌",class_name:"python 快乐学习班",number:1001})
WriteResult({"nInserted": 1})
> db.python_class.find()
{"_id": ObjectId("5c1b95f18ef102160306167a"), "name": "小萌", "class_name": "Python 学习班", "number": 1001 }
{"_id": ObjectId("5c1b9a298ef102160306167d"), "name": "小萌", "class_name": "Python 快乐学习班", "number": 1001 }
```

删除 python_class 集合中的所有文档，操作如下：

```
> db.python_class.remove({})
WriteResult({ "nRemoved" : 2 })
> db.python_class.find()
>
```

由打印结果可知，python_class 集合中已经没有文档了。

如果只想删除第一条找到的文档，则可以设置 justOne 为 1，操作如下：

```
>db.COLLECTION_NAME.remove(DELETION_CRITERIA,1)
```

在 mongo_use 数据库的 python_class 集合中插入两条 name 为"小萌"的文档，操作如下：

```
> db.python_class.insert({name: "小萌", class_name: "Python 快乐学习班", number: 1001})
WriteResult({"nInserted": 1})
> db.python_class.insert({name: "小萌", class_name: "Python 快乐学习班", number: 1001})
WriteResult({"nInserted": 1})
> db.python_class.find()                                                      :1001}
{"_id": ObjectId("5c1b9bcc8ef102160306167e"), "name": "小萌", "class_name": "Python 快乐学习班", "number": 1001 }
{"_id": ObjectId("5c1b9bd08ef102160306167f"), "name": "小萌", "class_name": "Python 快乐学习班", "number": 1001 }
```

删除第一条找到的文档，操作如下：

```
> db.python_class.remove({name:"小萌"}, 1)
WriteResult({"nRemoved": 1})
> db.python_class.find()
{"_id": ObjectId("5c1b9bd08ef102160306167f"), "name": "小萌", "class_ name": "Python 快乐学习班", "number": 1001}
```

由操作结果可知，删除了第一条找到的文档。

6.4.8 MongoDB 查询文档

MongoDB 查询文档使用 find()方法。find()方法以非结构化的方式来显示所有文档,语法格式如下:

```
> db.collection.find(query, projection)
```

参数说明如下。

query(可选):使用查询操作符指定查询条件。

projection(可选):使用投影操作符指定返回的键。查询时返回文档中所有键值,只需省略该参数即可(默认省略)。

如果需要以易读的方式来读取数据,可以使用 pretty()方法,语法格式如下:

```
>db.col.find().pretty()
```

pretty()方法以格式化的方式来显示所有文档。

如查询 python_class 集合中的数据,并以格式化方式显示,操作如下:

```
> db.python_class.find().pretty()
{
        "_id" : ObjectId("5c1b9bd08ef102160306167f"),
        "name" : "小萌",
        "class_name" : "Python 快乐学习班",
        "number" : 1001
}
```

除了 find()方法,还有一个 findOne()方法,它只返回一个文档。

若熟悉 SQL 语句,通过表 6-4 可以更好地理解 MongoDB 的条件语句查询。

表 6-4 MongoDB 与 SQL 类似语句比较

操作	格式	范例	SQL 类似语句
等于	{<key>:<value>}	db.python_class.find({"name":"小萌"}).pretty()	WHERE name='小萌'
小于	{<key>:{$lt:<value>}}	db.python_class.find({"number":{$lt:1000}}).pretty()	WHERE number < 1000
小于或等于	{<key>:{$lte:<value>}}	db.python_class.find({"number":{$lte:1000}}).pretty()	WHERE number <= 1000
大于	{<key>:{$gt:<value>}}	db.python_class.find({"number":{$gt:1000}}).pretty()	WHERE number > 1000
大于或等于	{<key>:{$gte:<value>}}	db.python_class.find({"number":{$gte:1000}}).pretty()	WHERE number >= 1000
不等于	{<key>:{$ne:<value>}}	db.python_class.find({"number":{$ne:1000}}).pretty()	WHERE number != 1000

接下来了解文档查询中的几个常见操作:AND 条件、OR 条件。

1. AND 条件

MongoDB 的 find()方法可以传入多个键(key),每个键(key)以逗号隔开,即常规 SQL 的 AND 条件。语法格式如下:

```
db.python_class.find({key1:value1, key2:value2}).pretty()
```

如查找集合 python_class 中 name 为"小萌"、number 为 1001 的文档,操作如下:

```
> db.python_class.find({name:"小萌", number:1001}).pretty()
{
    "_id" : ObjectId("5c1b9bd08ef102160306167f"),
    "name" : "小萌",
```

```
        "class_name" : "Python 快乐学习班",
        "number" : 1001
    }
```

该实例类似 SQL 中的如下 WHERE 语句：

```
WHERE name='小萌' AND number=1001
```

2. OR 条件

MongoDB 中的 OR 条件语句使用了关键字$or，语法格式如下：

```
>db.col.find(
    {
        $or: [
            {key1: value1}, {key2:value2}
        ]
    }
).pretty()
```

如查找集合 python_class 中 name 为"小明"或 number 为 1001 的文档，操作如下：

```
> db.python_class.find({$or:[{name:"小明"}, {number:1001}]}).pretty()
{
    "_id" : ObjectId("5c1b9bd08ef102160306167f"),
    "name" : "小萌",
    "class_name" : "Python 快乐学习班",
    "number" : 1001
}
{
    "_id" : ObjectId("5c1ba7588ef1021603061680"),
    "name" : "小明",
    "class_name" : "Python 快乐学习班",
    "number" : 1002
}
```

3. AND 和 OR 联合使用

例如：

```
> db.python_class.find({number:{$gt:1000}, $or:[{name:"小明"}, {name:"小萌"}]}).pretty()
{
    "_id" : ObjectId("5c1b9bd08ef102160306167f"),
    "name" : "小萌",
    "class_name" : "Python 快乐学习班",
    "number" : 1001
}
{
    "_id" : ObjectId("5c1ba7588ef1021603061680"),
    "name" : "小明",
    "class_name" : "Python 快乐学习班",
    "number" : 1002
}
```

该实例演示了 AND 和 OR 的联合使用方式，类似常规 SQL 语句：

```
WHERE number > 1000 AND (name = '小明' OR name = '小萌')'
```

6.4.9 条件操作符

MongoDB 中的条件操作符如下。
① 大于（>）：$gt。
② 小于（<）：$lt。
③ 大于等于（>=）：$gte。
④ 小于等于（<=）：$lte。

为了便于本节实例操作，mongo_use 数据库的 python_class 集合中插入一个文档，操作如下：

```
> db.python_class.insert({name:"小张", class_name:"Python 快乐学习班", number:1003})
WriteResult({"nInserted" : 1})
```

使用 find() 命令查看 python_class 集合中的数据：

```
> db.python_class.find().pretty()
{
    "_id" : ObjectId("5c1b9bd08ef102160306167f"),
    "name" : "小萌",
    "class_name" : "Python 快乐学习班",
    "number" : 1001
}
{
    "_id" : ObjectId("5c1ba7588ef1021603061680"),
    "name" : "小明",
    "class_name" : "Python 快乐学习班",
    "number" : 1002
}
{
    "_id" : ObjectId("5c1bad898ef1021603061681"),
    "name" : "小张",
    "class_name" : "Python 快乐学习班",
    "number" : 1003
}
```

1. MongoDB 大于操作符 $gt 的使用

获取 python_class 集合中 number 大于 1002 的数据，可以使用以下命令：

```
> db.python_class.find({number:{$gt:1002}}).pretty()
{
    "_id" : ObjectId("5c1bad898ef1021603061681"),
    "name" : "小张",
    "class_name" : "Python 快乐学习班",
    "number" : 1003
}
```

以上语句类似如下 SQL 语句：

```
SELECT * FROM python_class WHERE number > 1002
```

2. MongoDB 小于操作符 $lt 的使用

获取 python_class 集合中 number 小于 1002 的数据，可以使用以下命令：

```
> db.python_class.find({number:{$lt:1002}}).pretty()
{
    "_id" : ObjectId("5c1b9bd08ef102160306167f"),
    "name" : "小萌",
    "class_name" : "Python 快乐学习班",
    "number" : 1001
}
```

类似如下 SQL 语句：

```
SELECT * FROM python_class WHERE number < 1002
```

3. MongoDB 大于等于操作符 $gte 的使用

获取 python_class 集合中 number≥1002 的数据，可以使用以下命令：

```
> db.python_class.find({number:{$gte:1002}}).pretty()
{
    "_id" : ObjectId("5c1ba7588ef1021603061680"),
    "name" : "小明",
    "class_name" : "Python 快乐学习班",
    "number" : 1002
}
{
    "_id" : ObjectId("5c1bad898ef1021603061681"),
    "name" : "小张",
    "class_name" : "Python 快乐学习班",
    "number" : 1003
}
```

类似如下 SQL 语句：

```
SELECT * FROM python_class WHERE number >= 1002
```

4. MongoDB 小于等于操作符 $lte 的使用

获取 python_class 集合中 number≤1002 的数据，可以使用以下命令：

```
> db.python_class.find({number:{$lte:1002}}).pretty()
{
    "_id" : ObjectId("5c1b9bd08ef102160306167f"),
    "name" : "小萌",
    "class_name" : "Python 快乐学习班",
    "number" : 1001
}
{
    "_id" : ObjectId("5c1ba7588ef1021603061680"),
    "name" : "小明",
    "class_name" : "Python 快乐学习班",
    "number" : 1002
}
```

类似如下 SQL 语句：

```
SELECT * FROM python_class WHERE number <= 1002
```

6.4.10 $type 操作符

MongoDB 的条件操作符$type 是基于 BSON 类型来检索集合中匹配的数据类型，并返回结果。MongoDB 中可以使用的类型如表 6-5 所示。

表 6-5 MongoDB 中可以使用的类型

类 型	数 字	备 注
Double	1	
String	2	
Object	3	
Array	4	
Binary data	5	
Undefined	6	已废弃
Object ID	7	
Boolean	8	
Date	9	
Null	10	
Regular Expression	11	
JavaScript	13	
Symbol	14	
JavaScript (with scope)	15	
32-bit integer	16	
Timestamp	17	
64-bit integer	18	
Min key	255	Query with −1
Max key	127	

操作 mongo_use 数据库中的 python_class 集合，要获取 python_class 集合中 number 为 Double 的数据，操作如下：

```
db.python_class.find({number:{$type:1}}).pretty()
```

或

```
db.python_class.find({number:{$type:'double'}}).pretty()
```

使用以上操作方式，执行及结果如下：

```
> db.python_class.find({number:{$type:1}}).pretty()
{
    "_id" : ObjectId("5c1b9bd08ef102160306167f"),
    "name" : "小萌",
    "class_name" : "Python 快乐学习班",
    "number" : 1001
}
{
    "_id" : ObjectId("5c1ba7588ef1021603061680"),
    "name" : "小明",
```

```
        "class_name" : "Python 快乐学习班",
        "number" : 1002
    }
    {
        "_id" : ObjectId("5c1bad898ef1021603061681"),
        "name" : "小张",
        "class_name" : "Python 快乐学习班",
        "number" : 1003
    }
```

6.4.11 limit()和 skip()方法

1. limit()方法

如果需要在 MongoDB 中读取指定数量的数据记录，则可以使用 MongoDB 的 limit()方法。limit()方法接受一个数字参数，该参数指定从 MongoDB 中读取的记录条数。

limit()方法的语法格式如下：

```
db.COLLECTION_NAME.find().limit(NUMBER)
```

例如：

```
> db.python_class.find({},{number:1,_id:0}).limit(1)
{"number" : 1001}
> db.python_class.find({},{number:1,_id:0}).limit(2)
{"number" : 1001}
{"number" : 1002}
> db.python_class.find({},{number:1,_id:0}).limit()
{"number" : 1001}
{"number" : 1002}
{"number" : 1003}
```

由打印结果可知，如果没有指定 limit()方法中的参数，则显示集合中的所有数据。

2. skip()方法

除了可以使用 limit()方法来读取指定数量的数据，还可以使用 skip()方法跳过指定数量的数据。skip()方法同样接受一个数字参数作为跳过的记录条数。

skip()方法的语法格式如下：

```
db.COLLECTION_NAME.find().limit(NUMBER).skip(NUMBER)
```

例如：

```
> db.python_class.find({},{number:1, _id:0}).limit(1).skip(0)
{ "number" : 1001 }
> db.python_class.find({},{number:1, _id:0}).limit(1).skip(1)
{ "number" : 1002 }
> db.python_class.find({},{number:1, _id:0}).limit(2).skip(0)
{ "number" : 1001 }
{ "number" : 1002 }
> db.python_class.find({},{number:1, _id:0}).limit(2).skip(1)
{ "number" : 1002 }
{ "number" : 1003 }
> db.python_class.find({},{number:1, _id:0}).limit(2).skip(2)
```

```
{ "number" : 1003 }
> db.python_class.find({},{number:1, _id:0}).limit().skip()
{ "number" : 1001 }
{ "number" : 1002 }
{ "number" : 1003 }
> db.python_class.find({},{number:1, _id:0}).limit().skip(1)
{ "number" : 1002 }
{ "number" : 1003 }
> db.python_class.find({},{number:1, _id:0}).limit().skip(2)
{ "number" : 1003 }
> db.python_class.find({},{number:1, _id:0}).limit().skip(3)
>
```

由打印结果可知，skip()方法默认参数为 0，数字代表跳过指定数量的数据。

6.4.12 排序

MongoDB 中使用 sort()方法对数据进行排序。sort()方法的语法格式如下：

```
db.COLLECTION_NAME.find().sort({KEY:1})
```

sort()方法可以通过参数指定排序的字段，并使用 1 和-1 来指定排序的方式，其中 1 为升序排列，-1 为降序排列。

对 python_class 集合中的 number 字段分别进行升序和降序排序，实现方式如下：

```
> db.python_class.find({},{number:1, _id:0}).sort({number:1})
{ "number" : 1001 }
{ "number" : 1002 }
{ "number" : 1003 }
> db.python_class.find({},{number:1, _id:0}).sort({number:-1})
{ "number" : 1003 }
{ "number" : 1002 }
{ "number" : 1001 }
```

6.4.13 索引

MongoDB 使用 createIndex()方法来创建索引，其语法格式如下：

```
db.collection.createIndex(keys, options)
```

其中，keys 为 key-value 对，key 值为要创建的索引字段，value 取值 1 或-1，1 为指定按升序创建索引，如果按降序创建索引，则指定为-1。

索引通常能够极大地提高查询效率。如果没有索引，则 MongoDB 在读取数据时必须扫描集合中的每个文件并选取那些符合查询条件的记录。这种扫描全集合的查询效率是非常低的，特别在处理大量的数据时，查询要花费几十秒甚至几分钟，这对性能有要求的服务是非常致命的。

索引是特殊的数据结构，存储在一个易于遍历读取的数据集合中。索引是对数据库表中一列或多列的值进行排序的一种结构。

如对 python_class 集合中的 name 字段按升序创建索引，操作如下：

```
> db.python_class.createIndex({name:1})
{
    "createdCollectionAutomatically" : false,
    "numIndexesBefore" : 1,
    "numIndexesAfter" : 2,
    "ok" : 1
}
```

createIndex()方法也可以对多个字段创建索引（如关系型数据库中的复合索引）。

如对 python_class 集合中的 name 字段按升序、number 字段按降序创建索引，操作如下：

```
> db.python_class.createIndex({name:1, number:-1})
{
    "createdCollectionAutomatically" : false,
    "numIndexesBefore" : 2,
    "numIndexesAfter" : 3,
    "ok" : 1
}
```

createIndex()方法中的 options 为可选参数，其可选值如表 6-6 所示。

表 6-6 options 可选值

参数	类型	描述
background	Boolean	建索引过程会阻塞其他数据库操作，background 可指定以后台方式创建索引，即增加 background 可选参数。默认为 false
unique	Boolean	建立的索引是否唯一，若指定为 true，则创建唯一索引。默认为 false
name	string	索引的名称。如果未指定，则 MongoDB 通过连接索引的字段名和排序顺序生成一个索引名称
dropDups	Boolean	MongoDB 3.0 版本已废弃。在建立唯一索引时是否删除重复记录，指定为 true，则创建唯一索引。默认为 false
sparse	Boolean	对文档中不存在的字段数据不启用索引。如果设置为 true，则在索引字段中不会查询出不包含对应字段的文档。默认为 false
expireAfterSeconds	integer	指定一个以秒为单位的数值，完成 TTL 设定，设定集合的生存时间
v	index version	索引的版本号。默认的索引版本取决于 mongod 创建索引时运行的版本
weights	document	索引权重值，为 1~99999，表示该索引相对于其他索引字段的得分权重
default_language	string	对于文本索引，该参数决定了停用词及词干和词器的规则列表，默认为英语
language_override	string	对于文本索引，该参数指定了包含在文档中的字段名，语言覆盖默认值 language

如对 python_class 集合中的 class_name 和 number 字段在后台创建索引，操作如下：

```
> db.python_class.createIndex({class_name:1,number:-1},{background:true})
{
    "createdCollectionAutomatically" : false,
    "numIndexesBefore" : 3,
    "numIndexesAfter" : 4,
    "ok" : 1
}
```

即通过在创建索引时加 background:true 选项，让创建工作在后台执行。

6.4.14 聚合

MongoDB 中聚合（aggregate）的方法使用 aggregate()，其语法格式如下：

```
db.COLLECTION_NAME.aggregate(AGGREGATE_OPERATION)
```

聚合主要用于处理数据（诸如统计平均值、求和等），并返回计算后的数据结果，类似 SQL 语句中的 COUNT。

向 mongo_use 数据库的 python_class 集合中增加一个 name 为"小张"的文档，操作如下：

```
> db.python_class.insert({name:"小张", class_name:"Python 快乐学习班", number:1004})
WriteResult({"nInserted" : 1 })
```

查看 python_class 集合中的所有文档：

```
> db.python_class.find().pretty()
{
    "_id" : ObjectId("5c1b6e00b9444ea171d76ba8"),
    "name" : "小萌",
    "class_name" : "Python 快乐学习班",
    "number" : 1001
}
{
    "_id" : ObjectId("5c1b6f6bb9444ea171d76ba9"),
    "name" : "小明",
    "class_name" : "Python 快乐学习班",
    "number" : 1002
}
{
    "_id" : ObjectId("5c1c4363b9444ea171d76baa"),
    "name" : "小张",
    "class_name" : "Python 快乐学习班",
    "number" : 1003
}
{
    "_id" : ObjectId("5c1c4e1db9444ea171d76bab"),
    "name" : "小张",
    "class_name" : "Python 快乐学习班",
    "number" : 1004
}
```

由打印结果可知，python_class 集合中已经有 4 个文档，并且有 2 个 name 为"小张"的文档。

现在通过聚合方法计算各 name 值的个数，使用 aggregate()计算结果如下：

```
> db.python_class.aggregate([{$group:{_id:"$name",num_tutorial:{$sum:1}}}])
{ "_id" : "小明", "num_tutorial" : 1 }
{ "_id" : "小张", "num_tutorial" : 2 }
{ "_id" : "小萌", "num_tutorial" : 1 }
```

由计算及打印结果可知，使用 aggregate()方法得到了对应的结果集。

以上实例类似如下 SQL 语句：

SELECT name, count(*) FROM python_class GROUP BY NAME

通过 name 字段对数据进行分组，并计算 name 字段值相同的总数。

6.5 小结

本章主要讲解了 MongoDB 的一些基本知识，包括 MongoDB 简介、MongoDB 安装及其基本操作。本章所讲解的 MongoDB 内容都是比较基础的，并没有讲解 MongoDB 的高级内容。MongoDB 的高级内容包括分片、监控、数据库引用、覆盖索引查询、查询分析、高级索引、全文检索、正则表达式等。读者在掌握了基础知识后，可以借助网络等方式自学。本书由于篇幅限制，不做深入讲解。

6.6 实战演练

1. 在本地搭建 MongoDB 环境。
2. 根据书中示例，创建一个 MongoDB 数据库。
3. 对创建的示例，在数据库中创建集合、删除集合。
4. 对创建的示例，执行文档插入、文档更新、文档删除、文档查询等操作。
5. 结合示例，自行实现对 MongoDB 的排序、索引、聚合等操作。

第 7 章　Python 操作 MongoDB

本章将介绍如何通过 Python 代码操作 MongoDB。

数字校园的 mongo 山并不好维护，经常发动学生去拔除和修剪也不是长久之计，于是学校请了专业的植物护理人员来打理 mongo 山，需要修剪某些树木或拔除某些植物时，学校只需通知护理员，由他们具体操作即可。

在 Python 操作 MongoDB 的过程中，pymongo 充当了护理员的角色。

7.1　pymongo 安装

Python 要连接 MongoDB 需要驱动，一般使用 pymongo 驱动来连接。pymongo 是操作 MongoDB 的 Python 模块，是第三方模块，需要手动安装。

安装 pymongo 模块比较简单，可以直接使用 pip 安装，安装命令如下：

```
pip install pymongo
```

若提示找不到 pip，可以使用 pip3，根据本地环境中安装的 pip 情况使用即可。

也可以使用如下方式进行安装：

```
python -m pip install pymongo
```

若提示找不到 Python，可以使用 Python 3，根据本地环境中安装的 Python 使用即可。

也可以指定安装的版本，如下所示：

```
python -m pip install pymongo==3.7.2
```

更新 pymongo 命令如下：

```
python -m pip install --upgrade pymongo
```

执行安装命令后，可以通过指令模式输入如下命令，查看 pymongo 安装是否成功：

```
>>> import pymongo
>>>
```

若不报错，则安装成功。

或者可以编写一个 PY 文件，在 PY 文件中写一个 import 语句，执行 PY 文件，若不报错，则表示安装成功。操作示例如下（demo_import_pymongo.py）：

```
import pymongo
```

7.2　Python 连接 MongoDB

导入 pymongo 模块后，就可以用 Python 通过 pymongo 模块连接 MongoDB 了。连接 MongoDB 需要使用 pymongo 模块中的 MongoClient 对象，并且指定连接的 URL 地址和要连接或创建的数据库名。

Python 通过 pymongo 模块连接 MongoDB 的语法格式如下：

```
pymongo.MongoClient("mongodb://IP 地址:端口号/")
```

其中,"IP 地址"指的是 MongoDB 安装的服务器地址,若是本地,则可以使用 localhost;端口号是 MongoDB 对外提供的端口号,默认是 27017。

如本地连接 MongoDB,编写代码如下（conn_exp_1.py）：

```
import pymongo

pymongo.MongoClient("mongodb://localhost:27017/")
```

如果数据库设置了用户验证,那么需要在连接命令中加上验证信息。加上验证信息后的代码如下（conn_exp_2.py）：

```
import pymongo

pymongo.MongoClient("mongodb://username:password@localhost:27017/")
```

通过 pymongo 模块中的 MongoClient 对象连接 MongoDB,该连接会返回一个值,返回值中包含了所连接地址的所有数据库对象。

如在第 6 章的本地 MongoDB 中创建了一个 mongo_use 数据库,查看本地 MongoDB 中的所有数据库,操作如下：

```
> show dbs
admin      0.000GB
config     0.000GB
local      0.000GB
mongo_use  0.000GB
```

由打印结果可知,本地当前 MongoDB 服务器中有 4 个数据库,分别为 admin、config、local 和 mongo_use。

现需要通过编写 Python 代码打印 MongoDB 服务器中的所有数据库名。示例代码如下（conn_exp_3.py）：

```
import pymongo

mongo_client = pymongo.MongoClient("mongodb://localhost:27017/")
db_list = mongo_client.list_database_names()
for item in db_list:
    print('本地 MongoDB 服务器中的数据库: {}'.format(item))
```

执行以上代码,得到如下打印结果：

```
本地 MongoDB 服务器中的数据库: admin
本地 MongoDB 服务器中的数据库: config
本地 MongoDB 服务器中的数据库: local
本地 MongoDB 服务器中的数据库: mongo_use
```

由打印结果可知,示例代码将本地 MongoDB 服务器中的所有数据库名都打印了,得到的数据库个数和数据库名称与通过 MongoDB 命令得到的结果一致。

7.3 Python 对 MongoDB 的基本操作

与 MongoDB 的基础操作一样,Python 也可以对 MongoDB 进行创建数据库、创建集合、插入文档、更改文档、查询文档、删除文档等基本操作。

7.3.1 创建数据库

与 MongoDB 中创建数据库一样，通过 Python 操作 MongoDB 创建数据库时，数据库只有在内容插入后才会创建，即数据库创建后，要创建集合（数据表）并插入一个文档（记录），数据库才会真正创建。

继续第 6 章的操作，通过 Python 代码访问 mongo_use 数据库。操作示例如下（create_db_exp_1.py）：

```python
import pymongo

# mongo client
mongo_client = pymongo.MongoClient("mongodb://localhost:27017/")
# 数据库连接，若存在 mongo_use，则直接连接，否则创建
mongo_conn = mongo_client["mongo_use"]
print(mongo_conn)
```

执行该 PY 文件，得到输出结果如下：

```
Database(MongoClient(host=['localhost:27017'], document_class=dict, tz_aware=False, connect=True), 'mongo_use')
```

创建数据库时，若不确定某个数据库是否已经真正创建，则可以通过 MongoClient 对象中的 list_database_names()方法取得所有数据库名。

list_database_names()方法返回一个数据库名列表，可以从返回列表中查找对应数据库名是否已经存在。操作示例如下（create_db_exp_2.py）：

```python
import pymongo

# mongo client

mongo_client = pymongo.MongoClient("mongodb://localhost:27017/")
# 取得所有数据库名称
db_list = mongo_client.list_database_names()
if 'mongo_use' in db_list:
    print('database ({}) is exist.'.format('mongo_use'))
```

执行该 PY 文件，执行结果如下：

```
database (mongo_use) is exist.
```

7.3.2 创建集合

MongoDB 使用数据库对象来创建集合，即创建集合之前，需要先获取数据库对象。

使用 Python 代码创建集合的示例如下（create_col_exp_1.py）：

```python
import pymongo

# mongo client
mongo_client = pymongo.MongoClient("mongodb://localhost:27017/")
# 数据库连接，若存在 mongo_use，则直接连接，否则创建
mongo_conn = mongo_client["mongo_use"]
# 集合获取，若存在 python_class 集合，则直接返回，否则创建
mongo_col = mongo_conn["python_class"]
```

```
print(mongo_col)
```

执行该 PY 文件，得到结果如下：

```
Collection(Database(MongoClient(host=['localhost:27017'], document_class=dict, tz_aware=False,
connect=True), 'mongo_use'), 'python_class')
```

在 MongoDB 中，集合只有在内容插入后才会创建，即创建集合（数据表）后要再插入一个文档（记录），集合才会真正创建。

创建集合时，若不确定某个集合是否已经真正创建，则可以通过数据库对象中的 list_collection_names()方法取得所有集合名。list_collection_names()方法返回一个集合名列表，可以从返回列表中查找对应集合名是否已经存在。操作示例如下（create_col_exp_2.py）：

```
import pymongo

# mongo client
mongo_client = pymongo.MongoClient("mongodb://localhost:27017/")
# 取得数据库
mongo_conn = mongo_client["mongo_use"]
# 取得所有数据库名称
col_list = mongo_conn.list_collection_names()
if 'python_class' in col_list:
    print('collection ({}) is exist.'.format('python_class'))
```

执行该 PY 文件，得到如下执行结果：

```
collection (python_class) is exist.
```

7.3.3 查询文档

由第 6 章可知，MongoDB 中使用 find()和 find_one()方法查询集合中的数据，类似 SQL 中的 SELECT 语句。

pymongo 也提供了 find_one()方法，用于从 MongoDB 中查询一条数据。如需从 mongo_use 数据库的 python_class 集合中查询一个文档，操作方法如下（query_exp_1.py）：

```
import pymongo

# mongo client
mongo_client = pymongo.MongoClient("mongodb://localhost:27017/")
# 数据库连接，若存在 mongo_use，则直接连接，否则创建
mongo_conn = mongo_client["mongo_use"]
# 集合获取，若存在 python_class 集合，则直接返回，否则创建
mongo_col = mongo_conn["python_class"]
# 查询集合中的一条数据
query_one = mongo_col.find_one()
print(query_one)
```

执行该 PY 文件，得到执行结果如下：

```
{'_id': ObjectId('5c1b9bd08ef102160306167f'), 'name': '小萌', 'class_name': 'Python 快乐学习
班', 'number': 1001.0}
```

该示例只能查询一个文档，若查询全部文档，需要使用 pymongo 提供的 find()方法。如从 python_class 集合中查询所有文档，则操作方法如下（query_exp_2.py）：

```python
import pymongo

# mongo client
mongo_client = pymongo.MongoClient("mongodb://localhost:27017/")
# 数据库连接，若存在 mongo_use，则直接连接，否则创建
mongo_conn = mongo_client["mongo_use"]
# 集合获取，若存在 python_class 集合，则直接返回，否则创建
mongo_col = mongo_conn["python_class"]
# 查询集合中所有数据
query_all = mongo_col.find()
for item in query_all:
    print(item)
```

执行该 PY 文件，得到执行结果如下：

```
{'_id': ObjectId('5c1b9bd08ef102160306167f'), 'name':'小萌', 'class_name': 'Python 快乐学习班', 'number': 1001.0}
{'_id': ObjectId('5c1ba7588ef1021603061680'), 'name':'小明', 'class_name': 'Python 快乐学习班', 'number': 1002.0}
{'_id': ObjectId('5c1ce74a8ef1021603061684'), 'name':'小张', 'class_name': 'Python 快乐学习班', 'number': 1003.0}
{'_id': ObjectId('5c1ce7518ef1021603061685'), 'name':'小张', 'class_name': 'Python 快乐学习班', 'number': 1004.0}
```

pymongo 提供的 find() 方法还可以查询指定字段的数据，将要返回的字段对应值设置为 1 即可。

如从 python_class 集合中查询 name 和 number 字段，操作方法如下（query_exp_3.py）：

```python
import pymongo

# mongo client
mongo_client = pymongo.MongoClient("mongodb://localhost:27017/")
# 数据库连接，若存在 mongo_use，则直接连接，否则创建
mongo_conn = mongo_client["mongo_use"]
# 集合获取，若存在 python_class 集合，则直接返回，否则创建
mongo_col = mongo_conn["python_class"]
# 查询指定字段的数据
query_all = mongo_col.find({}, {"_id": 0, "name": 1, "number": 1})
for item in query_all:
    print(item)
```

执行该 PY 文件，得到执行结果如下：

```
{'name': '小萌', 'number': 1001.0}
{'name': '小明', 'number': 1002.0}
{'name': '小张', 'number': 1003.0}
{'name': '小张', 'number': 1004.0}
```

注意：除了 _id 字段，不能在一个对象中同时指定 0 和 1。也就是说，如果设置了一个字段为 1，不能设置另一个字段为 0；设置了一个字段为 0，不能设置另一个字段为 1。除了 _id 字段，其他字段要么都设置为 1，要么都设置为 0。

如以下示例中，除了 class_name 字段，其他字段都返回值（query_exp_4.py）：

```python
import pymongo
```

```
# mongo client
mongo_client = pymongo.MongoClient("mongodb://localhost:27017/")
# 数据库连接,若存在 mongo_use,则直接连接,否则创建
mongo_conn = mongo_client["mongo_use"]
# 集合获取,若存在 python_class 集合,则直接返回,否则创建
mongo_col = mongo_conn["python_class"]
# 查询指定字段的数据
query_all = mongo_col.find({}, {"class_name": 0})

for item in query_all:
    print(item)
```

执行该 PY 文件,得到执行结果如下:

```
{'_id': ObjectId('5c1b9bd08ef102160306167f'), 'name': '小萌', 'number': 1001.0}
{'_id': ObjectId('5c1ba7588ef1021603061680'), 'name': '小明', 'number': 1002.0}
{'_id': ObjectId('5c1ce74a8ef1021603061684'), 'name': '小张', 'number': 1003.0}
{'_id': ObjectId('5c1ce7518ef1021603061685'), 'name': '小张', 'number': 1004.0}
```

与其他查询有过滤功能一样,pymongo 也提供了过滤查询,可以在 find() 中设置参数来过滤数据。

如从 python_class 集合中查询 name 为 "小明" 的文档,操作如下(query_exp_2.py):

```
import pymongo

# mongo client
mongo_client = pymongo.MongoClient("mongodb://localhost:27017/")
# 数据库连接,若存在 mongo_use,则直接连接,否则创建
mongo_conn = mongo_client["mongo_use"]
# 集合获取,若存在 python_class 集合,则直接返回,否则创建
mongo_col = mongo_conn["python_class"]
# 根据指定条件查询
query_all = mongo_col.find({"name": "小明"})
for item in query_all:
    print(item)
```

执行该 PY 文件,得到执行结果如下:

```
{'_id': ObjectId('5c1ba7588ef1021603061680'), 'name':'小明', 'class_name': 'Python快乐学习班', 'number': 1002.0}
```

查询的条件语句中还可以使用修饰符。如从 python_class 集合中读取 number 字段值大于 1002 的数据,修饰符条件为 {"$gt": 1002},示例如下(query_exp_6.py):

```
import pymongo

# mongo client
mongo_client = pymongo.MongoClient("mongodb://localhost:27017/")
# 数据库连接,若存在 mongo_use,则直接连接,否则创建
mongo_conn = mongo_client["mongo_use"]
# 集合获取,若存在 python_class 集合,则直接返回,否则创建
mongo_col = mongo_conn["python_class"]
# 查询条件

condition_dict = {"number": {"$gt": 1002}}
# 高级查询
```

```
query_all = mongo_col.find(condition_dict)
for item in query_all:
    print(item)
```

执行该 PY 文件，得到执行结果如下：

```
{'_id': ObjectId('5c1ce74a8ef1021603061684'), 'name':'小张', 'class_name': 'Python 快乐学习班', 'number': 1003.0}
{'_id': ObjectId('5c1ce7518ef1021603061685'), 'name':'小张', 'class_name': 'Python 快乐学习班', 'number': 1004.0}
```

如果对查询结果设置返回指定条数的记录，那么可以使用 limit()方法，该方法只接受一个数字参数。以下示例用于从 python_class 集合中读取 2 条记录（query_exp_7.py）：

```
import pymongo

# mongo client
mongo_client = pymongo.MongoClient("mongodb://localhost:27017/")
# 数据库连接，若存在 mongo_use，则直接连接，否则创建
mongo_conn = mongo_client["mongo_use"]
# 集合获取，若存在 python_class 集合，则直接返回，否则创建
mongo_col = mongo_conn["python_class"]
# 返回指定条数记录
query_all = mongo_col.find().limit(2)
for item in query_all:
    print(item)
```

执行该 PY 文件，得到执行结果如下：

```
{'_id': ObjectId('5c1b9bd08ef102160306167f'), 'name':'小萌', 'class_name': 'Python 快乐学习班', 'number': 1001.0}
{'_id': ObjectId('5c1ba7588ef1021603061680'), 'name':'小明', 'class_name': 'Python 快乐学习班', 'number': 1002.0}
```

7.3.4 插入文档

插入文档使用 insert()方法。在 pymongo 中提供了 insert_one()方法向 MongoDB 中插入一个文档，该方法的第一个参数是字典 key-value 对。

如向 stu_info 集合中插入一个文档，示例如下（insert_exp_1.py）：

```
import pymongo

# mongo client
mongo_client = pymongo.MongoClient("mongodb://localhost:27017/")

# 数据库连接，若存在 mongo_use，则直接连接，否则创建
mongo_conn = mongo_client["mongo_use"]
# 集合获取，若存在 stu_info 集合，则直接返回，否则创建
col_name = mongo_conn["stu_info"]
# 插入值字典
insert_dict = {"name": "小强", "number": 1001, "email": "xiaoqiang@abc.com"}
# 集合中插入一个文档
result = col_name.insert_one(insert_dict)
```

```
print(result)
# 查询集合中所有文档
query_all = col_name.find()
for item in query_all:
    print(item)
```

执行该 PY 文件，得到执行结果如下：

```
<pymongo.results.InsertOneResult object at 0x0000000002BF3F48>
{'_id': ObjectId('5c1da41319eb3f14a047b6af'), 'name': '小强', 'number': 1001, 'email': 'xiaoqiang@abc.com'}
```

由打印结果可知，文档已成功插入 stu_info 集合。这里不要忘记，若集合不存在，会创建对应的集合。

insert_one()方法返回 InsertOneResult 对象，该对象包含 inserted_id 属性，它是插入文档的 id 值。如上面示例中需要查看插入文档的 id 值，可以把 print(result)修改如下：

```
print(result.inserted_id)
```

若插入多个文档，则需要使用 insert_many()方法，其第一个参数是字典列表。

如下示例用于向 stu_info 集合中插入多个文档，并打印所有 id 值（insert_exp_2.py）：

```
import pymongo

# mongo client
mongo_client = pymongo.MongoClient("mongodb://localhost:27017/")
# 数据库连接，若存在 mongo_use，则直接连接，否则创建
mongo_conn = mongo_client["mongo_use"]
# 集合获取，若存在 stu_info 集合，则直接返回，否则创建
col_name = mongo_conn["stu_info"]
# 插入值列表
insert_list = [
    {"name": "小王", "number": 1002, "phone": 15217639876},
    {"name": "小李", "number": 1003, "address": "beijing"},
    {"name": "小张", "number": 1004, "habit": "旅游，看书", "email": "xiaozhang@abc.com"},
    {"name": "小刘", "number": 1005, "qq": 123456, "email": "xiaoliu@123.com"}
]

# 集合中插入多个文档
result = col_name.insert_many(insert_list)
# 输出插入的所有文档对应的 id 值
print('所有文档id值：\n{}'.format(result.inserted_ids))
# 查询集合中所有文档
query_all = col_name.find()
print('stu_info 集合中的所有文档：')
for item in query_all:
    print(item)
```

执行该 PY 文件，得到执行结果如下。

所有文档 id 值如下：

```
[ObjectId('5c1da78e19eb3f27e028d166'), ObjectId('5c1da78e19eb3f27e028 d167'),
 ObjectId('5c1da78e19eb3f27e028d168'), ObjectId('5c1da78e19eb3f27e028d 169')]
```

stu_info 集合中的所有文档如下：

```
{'_id': ObjectId('5c1da78e19eb3f27e028d166'), 'name': '小王', 'number': 1002, 'phone': 15217639876}
{'_id': ObjectId('5c1da78e19eb3f27e028d167'), 'name': '小李', 'number': 1003, 'address': 'beijing'}
{'_id': ObjectId('5c1da78e19eb3f27e028d168'), 'name': '小张', 'number': 1004, 'habit': '旅游, 看书', 'email': 'xiaozhang@abc.com'}
{'_id': ObjectId('5c1da78e19eb3f27e028d169'), 'name': '小刘', 'number': 1005, 'qq': 123456, 'email': 'xiaoliu@123.com'}
```

由打印结果可知，多个文档已经成功插入集合中。

同时可知，一个集合中插入文档的字段名可以各不相同，字段个数也可以不同，这也是前面提到的 NoSQL 的便捷之处，不需要每个文档的字段名都相同，也不需要有相同的字段个数。

在前面的操作示例可知，_id 字段值比较长，是否可以将_id 设置成关系型数据库那样的，以数字形式展示呢？

pymongo 可以自己指定_id 的插入值。如在集合中插入指定_id 的文档，示例如下（insert_exp_3.py）：

```python
import pymongo

# mongo client
mongo_client = pymongo.MongoClient("mongodb://localhost:27017/")
# 数据库连接，若存在 mongo_use，则直接连接，否则创建
mongo_conn = mongo_client["mongo_use"]
# 集合获取，若存在 language_info 集合，则直接返回，否则创建
col_name = mongo_conn["language_info"]
# 插入值列表
insert_list = [
    {"_id": 1, "name": "Python", "web_site": "www.python.org"},
    {"_id": 2, "name": "BigData", "从业人员": 236783},
    {"_id": 3, "name": "AI", "人才需求": 500000},
]
# 插入指定 id 的多个文档
result = col_name.insert_many(insert_list)
# 输出插入的所有文档对应的 id 值
print('所有文档id值: \n{}'.format(result.inserted_ids))
# 查询集合中所有文档
query_all = col_name.find()
print('stu_info 集合中的所有文档: ')
for item in query_all:
    print(item)
```

执行该 PY 文件，得到执行结果如下。

所有文档的_id 值如下：

```
[1, 2, 3]
```

stu_info 集合中的所有文档如下：

```
{'_id': 1, 'name': 'Python', 'web_site': 'www.python.org'}
{'_id': 2, 'name': 'BigData', '从业人员': 236783}
{'_id': 3, 'name': 'AI', '人才需求': 500000}
```

7.3.5 更改文档

在 MongoDB 中，更改文档使用的是 update() 方法。在 pymongo 模块中使用 update_one() 方法修改文档中的记录，其第一个参数为查询的条件，第二个参数为要修改的字段。

如将 mongo_use 数据库中 python_class 集合的 number 为 1004 的文档的 name 更改为"小李"，操作如下（update_exp_1.py）：

```python
import pymongo

def get_col():
    """
    取得指定集合
    :return: 指定集合
    """
    # mongo client
    mongo_client = pymongo.MongoClient("mongodb://localhost:27017/")
    # 数据库连接，若存在 mongo_use，则直接连接，否则创建
    mongo_conn = mongo_client["mongo_use"]
    # 集合获取，若存在 python_class 集合，则直接返回，否则创建
    mongo_col = mongo_conn["python_class"]
    return mongo_col

def query_all():
    """
    查询集合中所有数据
    :return: None
    """
    # 查询集合中所有数据
    all_result = get_col().find()
    # 输出修改后的 python_class 集合
    for item in all_result:
        print(item)

def one_update():
    """
    更新一个文档
    :return: None
    """
    print('更改文档前，文档中的数据为：')
    query_all()
    # 集合获取，若存在 python_class 集合，则直接返回，否则创建
    mongo_col = get_col()
    # 查询条件
    condition_dict = {"number": 1004}
    # 新 name 值
    new_name = {"$set": {"name": "小李"}}
    # 更新
    obj = mongo_col.update_one(condition_dict, new_name)
    print("共修改{}个文档".format(obj.modified_count))
    print('更改文档后，文档中的数据为：')
    query_all()
```

```python
if __name__ == "__main__":
    one_update()
```

执行该 PY 文件，得到执行结果如下：

```
更改文档前，文档中的数据为：
    {'_id': ObjectId('5c1b9bd08ef102160306167f'), 'name':'小萌', 'class_name': 'Python 快乐学习班', 'number': 1001.0}
    {'_id': ObjectId('5c1ba7588ef1021603061680'), 'name':'小明', 'class_name': 'Python 快乐学习班', 'number': 1002.0}
    {'_id': ObjectId('5c1ce74a8ef1021603061684'), 'name':'小张', 'class_name': 'Python 快乐学习班', 'number': 1003.0}
    {'_id': ObjectId('5c1ce7518ef1021603061685'), 'name':'小张', 'class_name': 'Python 快乐学习班', 'number': 1004.0}
    共修改 1 个文档
    更改文档后，文档中的数据为：
    {'_id': ObjectId('5c1b9bd08ef102160306167f'), 'name':'小萌', 'class_name': 'Python 快乐学习班', 'number': 1001.0}
    {'_id': ObjectId('5c1ba7588ef1021603061680'), 'name':'小明', 'class_name': 'Python 快乐学习班', 'number': 1002.0}
    {'_id': ObjectId('5c1ce74a8ef1021603061684'), 'name':'小张', 'class_name': 'Python 快乐学习班', 'number': 1003.0}
    {'_id': ObjectId('5c1ce7518ef1021603061685'), 'name':'小李', 'class_name': 'Python 快乐学习班', 'number': 1004.0}
```

由打印结果可知，文档中 number 为 1004 的文档的 name 由"小张"更改为了"小李"。该示例代码做了代码的封装，将代码段抽象成了函数。

update_one()方法只能修改匹配到的第一条记录，如果修改所有匹配到的记录，那么可以使用 update_many()。

如将 python_class 集合的 class_name 为 "Python 快乐学习班" 的文档更改为 "Python 进阶班"，则操作如下（update_exp_2.py）：

```python
import pymongo

def get_col():
    """
    取得指定集合
    :return: 指定集合
    """
    # mongo client
    mongo_client = pymongo.MongoClient("mongodb://localhost:27017/")
    # 数据库连接，若存在 mongo_use，则直接连接，否则创建
    mongo_conn = mongo_client["mongo_use"]
    # 集合获取，若存在 python_class 集合，则直接返回，否则创建
    mongo_col = mongo_conn["python_class"]
    return mongo_col

def query_all():
    """
    查询集合中的所有数据
    :return: None
    """
```

```
        # 查询集合中的所有数据
        all_result = get_col().find()
        # 输出修改后的python_class集合
        for item in all_result:
            print(item)

    def update_multi():
        """
        更新多个文档
        :return: None
        """
        print('更改文档前，文档中的数据为：')
        query_all()
        # 集合获取，若存在python_class集合，则直接返回，否则创建
        mongo_col = get_col()
        # 查询条件
        condition_dict = {"class_name": "Python 快乐学习班"}
        # 新值
        new_value = {"$set": {"class_name": "Python 进阶班"}}
        # 更新
        obj = mongo_col.update_many(condition_dict, new_value)
        print("共修改{}个文档".format(obj.modified_count))
        print('更改文档后，文档中的数据为：')
        query_all()

    if __name__ == "__main__":
        update_multi()
```

执行该 PY 文件，得到执行结果如下：

```
更改文档前，文档中的数据为：
    {'_id': ObjectId('5c1b9bd08ef102160306167f'), 'name':'小萌', 'class_name': 'Python 快乐学习班', 'number': 1001.0}
    {'_id': ObjectId('5c1ba7588ef1021603061680'), 'name':'小明', 'class_name': 'Python 快乐学习班', 'number': 1002.0}
    {'_id': ObjectId('5c1ce74a8ef1021603061684'), 'name':'小张', 'class_name': 'Python 快乐学习班', 'number': 1003.0}
    {'_id': ObjectId('5c1ce7518ef1021603061685'), 'name':'小李', 'class_name': 'Python 快乐学习班', 'number': 1004.0}
    共修改 4 个文档
    更改文档后，文档中的数据为：
    {'_id': ObjectId('5c1b9bd08ef102160306167f'), 'name':'小萌', 'class_name': 'Python 进阶班', 'number': 1001.0}
    {'_id': ObjectId('5c1ba7588ef1021603061680'), 'name':'小明', 'class_name': 'Python 进阶班', 'number': 1002.0}
    {'_id': ObjectId('5c1ce74a8ef1021603061684'), 'name':'小张', 'class_name': 'Python 进阶班', 'number': 1003.0}
    {'_id': ObjectId('5c1ce7518ef1021603061685'), 'name':'小李', 'class_name': 'Python 进阶班', 'number': 1004.0}
```

由结果可知，文档中 class_name 为"Python 快乐学习班"的文档全部更改为了"Python 进阶班"。

7.3.6 文档排序

在 MongoDB 中使用 sort()方法对数据进行排序。pymongo 也是使用 sort()方法对数据进行排序。

sort()方法可以指定升序或降序排序。其第一个参数为要排序的字段，第二个字段指定排序规则，1 为升序，-1 为降序，默认为升序。

对 mongo_use 数据库 python_class 集合中所有文档根据 number 字段进行升序、降序排序，操作如下（sort_exp_1.py）：

```python
import pymongo

def get_col():
    """
    取得指定集合
    :return: 指定集合
    """
    # mongo client
    mongo_client = pymongo.MongoClient("mongodb://localhost:27017/")
    # 数据库连接，若存在 mongo_use 则直接连接，否则创建
    mongo_conn = mongo_client["mongo_use"]
    # 集合获取，若存在 python_class 集合，则直接返回，否则创建
    mongo_col = mongo_conn["python_class"]
    return mongo_col

def execute_sort():
    """
    执行排序
    :return: None
    """
    # 集合获取，若存在 python_class 集合，则直接返回，否则创建
    mongo_col = get_col()
    # 升序排序
    obj_asce_sort = mongo_col.find().sort("number")
    print('升序排序结果: ')
    for item in obj_asce_sort:
        print(item)

    # 降序排序
    obj_desc_sort = mongo_col.find().sort("number", -1)
    print('降序排序结果: ')
    for item in obj_desc_sort:
        print(item)

if __name__ == "__main__":
    execute_sort()
```

执行该 PY 文件，得到执行结果如下：

```
升序排序结果:
{'_id': ObjectId('5c1b9bd08ef102160306167f'), 'name': '小萌', 'class_name': 'Python 进阶班', 'number': 1001.0}
{'_id': ObjectId('5c1ba7588ef1021603061680'), 'name': '小明', 'class_name': 'Python 进阶班', 'number': 1002.0}
```

```
        {'_id': ObjectId('5c1ce74a8ef1021603061684'), 'name': '小张', 'class_name': 'Python 进阶班',
'number': 1003.0}
        {'_id': ObjectId('5c1ce7518ef1021603061685'), 'name': '小李', 'class_name': 'Python 进阶班',
'number': 1004.0}
        降序排序结果:
        {'_id': ObjectId('5c1ce7518ef1021603061685'), 'name': '小李', 'class_name': 'Python 进阶班',
'number': 1004.0}
        {'_id': ObjectId('5c1ce74a8ef1021603061684'), 'name': '小张', 'class_name': 'Python 进阶班',
'number': 1003.0}
        {'_id': ObjectId('5c1ba7588ef1021603061680'), 'name': '小明', 'class_name': 'Python 进阶班',
'number': 1002.0}
        {'_id': ObjectId('5c1b9bd08ef102160306167f'), 'name': '小萌', 'class_name': 'Python 进阶班',
'number': 1001.0}
```

除了单个字段的排序，也可以对多个字段进行排序；排序方式除了可以用数字 1 或 -1，也可以使用 pymongo.ASCENDING(1)或 pymongo.DESCENDING(-1)。

对 python_class 集合中所有文档根据 name 和 number 字段进行降序排序，操作如下（sort_exp_2.py）：

```python
import pymongo

def get_col():
    """
    取得指定集合
    :return: 指定集合
    """
    # mongo client
    mongo_client = pymongo.MongoClient("mongodb://localhost:27017/")
    # 数据库连接，若存在 mongo_use，则直接连接，否则创建
    mongo_conn = mongo_client["mongo_use"]
    # 集合获取，若存在 python_class 集合，则直接返回，否则创建
    mongo_col = mongo_conn["python_class"]
    return mongo_col

def execute_sort():
    """
    更新一个文档
    :return: None
    """
    # 集合获取，若存在 python_class 集合，则直接返回，否则创建
    mongo_col = get_col()
    # 排序
    obj_sort = mongo_col.find().sort([
        ("name", pymongo.DESCENDING),
        ("number", pymongo.DESCENDING)])
    for item in obj_sort:
        print(item)

if __name__ == "__main__":
    execute_sort()
```

执行该 PY 文件，得到执行结果如下：

```
        {'_id': ObjectId('5c1b9bd08ef102160306167f'), 'name': '小萌', 'class_name': 'Python 进阶班',
'number': 1001.0}
        {'_id': ObjectId('5c1ce7518ef1021603061685'), 'name': '小李', 'class_name': 'Python 进阶班',
'number': 1004.0}
        {'_id': ObjectId('5c1ba7588ef1021603061680'), 'name': '小明', 'class_name': 'Python 进阶班',
'number': 1002.0}
        {'_id': ObjectId('5c1ce74a8ef1021603061684'), 'name': '小张', 'class_name': 'Python 进阶班',
'number': 1003.0}
```

7.3.7 删除文档

pymongo 使用 delete_one()方法删除一个文档,其第一个参数为查询对象,指定要删除哪些数据。

以下示例删除 stu_info 集合中 name 字段值为"小王"的文档(delete_exp_1.py):

```
import pymongo

def get_col():
    """
    取得指定集合
    :return: 指定集合
    """
    # mongo client
    mongo_client = pymongo.MongoClient("mongodb://localhost:27017/")
    # 数据库连接,若存在 mongo_use,则直接连接,否则创建
    mongo_conn = mongo_client["mongo_use"]
    # 集合获取,若存在 python_class 集合,则直接返回,否则创建
    mongo_col = mongo_conn["stu_info"]

    return mongo_col

def delete_one_row():
    """
    删除一个文档
    :return: None
    """
    print('删除前: ')
    query_all()
    # 集合获取,若存在 python_class 集合,则直接返回,否则创建
    mongo_col = get_col()
    # 查询条件
    condition_dict = {"name": "小王"}
    mongo_col.delete_one(condition_dict)
    print('删除后: ')
    query_all()

def query_all():
    """
    查询集合中的所有数据
    :return: None
    """
    # 查询集合中的所有数据
```

```python
        all_result = get_col().find()
        # 输出修改后的 python_class 集合
        for item in all_result:
            print(item)

    if __name__ == "__main__":
        delete_one_row()
```

执行该 PY 文件，得到执行结果如下：

```
删除前：
    {'_id': ObjectId('5c1da78e19eb3f27e028d166'), 'name': '小王', 'number': 1002, 'phone': 15217639876}
    {'_id': ObjectId('5c1da78e19eb3f27e028d167'), 'name': '小李', 'number': 1003, 'address': 'beijing'}
    {'_id': ObjectId('5c1da78e19eb3f27e028d168'), 'name': '小张', 'number': 1004, 'habit': '旅游,看书', 'email': 'xiaozhang@abc.com'}
    {'_id': ObjectId('5c1da78e19eb3f27e028d169'), 'name': '小刘', 'number': 1005, 'qq': 123456, 'email': 'xiaoliu@123.com'}
删除后：
    {'_id': ObjectId('5c1da78e19eb3f27e028d167'), 'name': '小李', 'number': 1003, 'address': 'beijing'}

    {'_id': ObjectId('5c1da78e19eb3f27e028d168'), 'name': '小张', 'number': 1004, 'habit': '旅游,看书', 'email': 'xiaozhang@abc.com'}
    {'_id': ObjectId('5c1da78e19eb3f27e028d169'), 'name': '小刘', 'number': 1005, 'qq': 123456, 'email': 'xiaoliu@123.com'}
```

由结果可知，执行删除后，name 为"小王"的文档已经被删除。

delete_one()方法只能删除一个文档，要删除多个文档，可以使用 delete_many()方法，其第一个参数为查询对象，指定要删除哪些数据。

删除 stu_info 集合中 number 字段值大于 1003 的文档，操作如下（delete_exp_2.py）：

```python
import pymongo

def get_col():
    """
    取得指定集合
    :return: 指定集合
    """
    # mongo client
    mongo_client = pymongo.MongoClient("mongodb://localhost:27017/")
    # 数据库连接，若存在 mongo_use，则直接连接，否则创建
    mongo_conn = mongo_client["mongo_use"]
    # 集合获取，若存在 python_class 集合，则直接返回，否则创建
    mongo_col = mongo_conn["stu_info"]
    return mongo_col

def delete_many_row():
    """
    删除多个文档
    :return: None
    """
    print('删除前：')
```

```python
        query_all()
        # 集合获取，若存在 python_class 集合，则直接返回，否则创建
        mongo_col = get_col()
        # 查询条件
        condition_dict = {"number": {"$gt": 1003}}
        mongo_col.delete_many(condition_dict)
        print('删除后: ')
        query_all()

def query_all():
    """
    查询集合中的所有数据
    :return: None
    """
    # 查询集合中的所有数据
    all_result = get_col().find()
    # 输出修改后的 python_class 集合
    for item in all_result:
        print(item)

if __name__ == "__main__":
    delete_many_row()
```

执行该 PY 文件，得到执行结果如下：

```
删除前：
{'_id': ObjectId('5c1da78e19eb3f27e028d167'), 'name': '小李', 'number': 1003, 'address': 'beijing'}
{'_id': ObjectId('5c1da78e19eb3f27e028d168'), 'name': '小张', 'number': 1004,
 'habit': '旅游，看书', 'email': 'xiaozhang@abc.com'}
{'_id': ObjectId('5c1da78e19eb3f27e028d169'), 'name': '小刘', 'number': 1005, 'qq': 123456,
 'email': 'xiaoliu@123.com'}
删除后：
{'_id': ObjectId('5c1da78e19eb3f27e028d167'), 'name': '小李', 'number': 1003, 'address': 'beijing'}
```

在调用 delete_many()方法时，如果传入的是一个空的查询对象，则会删除集合中的所有文档。操作示例如下（delete_exp_3.py）：

```python
import pymongo

def get_col():
    """
    取得指定集合
    :return: 指定集合
    """
    # mongo client
    mongo_client = pymongo.MongoClient("mongodb://localhost:27017/")
    # 数据库连接，若存在 mongo_use，则直接连接，不存在则创建
    mongo_conn = mongo_client["mongo_use"]
    # 集合获取，若存在 python_class 集合，则直接返回，不存在则创建
    mongo_col = mongo_conn["stu_info"]
    return mongo_col

def delete_all_row():
    """
```

```
        删除所有文档
        :return: None
        """
        print('删除前: ')
        query_all()
        # 集合获取,若存在 python_class 集合,则直接返回,否则创建
        mongo_col = get_col()
        mongo_col.delete_many({})
        print('删除后: ')
        query_all()

    def query_all():
        """
        查询集合中的所有数据
        :return: None
        """
        # 查询集合中的所有数据
        all_result = get_col().find()
        # 输出修改后的 python_class 集合
        for item in all_result:
            print(item)

    if __name__ == "__main__":
        delete_all_row()
```

执行该 PY 文件,得到执行结果如下:

```
删除前:
{'_id': ObjectId('5c1da78e19eb3f27e028d167'), 'name': '小李', 'number': 1003, 'address': 'beijing'}
删除后:
```

pymongo 中删除一个集合也是使用 drop()方法。从 mongo_use 数据库中删除 stu_info 集合的操作如下(delete_exp_4.py):

```
import pymongo

def get_db():
    """
    取得指定数据库
    :return: 指定数据库
    """
    # mongo client
    mongo_client = pymongo.MongoClient("mongodb://localhost:27017/")
    # 数据库连接,若存在 mongo_use,则直接连接,否则创建
    mongo_conn = mongo_client["mongo_use"]
    return mongo_conn

def get_all_col():
    """
    打印所有集合名
    :return: None
    """
    mongo_col_obj = get_db()
    col_list = mongo_col_obj.list_collection_names()
```

```python
    for col in col_list:
        print('集合名：{}'.format(col))

def drop_col():
    """
    删除集合
    :return: None
    """
    print('删除集合 stu_info 前：')
    get_all_col()
    mongo_conn = get_db()
    mongo_col = mongo_conn["stu_info"]
    mongo_col.drop()
    print('删除集合 stu_info 后：')
    get_all_col()

if __name__ == "__main__":
    drop_col()
```

执行该 PY 文件，得到执行结果如下：

```
删除集合 stu_info 前：
集合名：python_class
集合名：language_info
集合名：stu_info
删除集合 stu_info 后：
集合名：python_class
集合名：language_info
```

由结果可知，集合 stu_info 已经被成功删除。

7.4 小结

本章内容主要是通过 pymongo 模块操作 MongoDB。本章内容比较偏重实际操作，并没有过多的理论性内容。要更好地掌握本章的内容，读者还需要多进行实际操作。

Python 通过 pymongo 模块操作 MongoDB 等 NoSQL 的过程与操作 MySQL 等关系型数据库类似，都是先获取数据库连接，再对数据库表或集合进行相关操作，具体操作方式有些不同而已。

Python 通过 pymongo 模块操作 MongoDB 等 NoSQL 的操作远不止本章所讲解的内容，本章只是基础讲解，更多的操作在遇到具体项目问题时，希望读者可以借助本章讲解的基础内容做出应变。

7.5 实战演练

1. 安装 pymongo 环境。
2. 使用 pymongo 创建数据库。
3. 对创建的数据库执行文档的创建、查询、更改、删除等操作。

第 8 章 文件读写

前面章节中主要讲解的是对关系型或非关系型数据库的操作处理,在实际应用中,除了对数据库的操作,经常需要操作文件,从文件中读取数据或将数据保存到文件中。本章将介绍如何从文件中读取数据,如何将数据写入文件中,如何将文件中读取的数据写入到数据库中,以及如何将数据库中读取的数据写入文件中等操作。

本章将讲解 TXT 文件读写、CSV 文件读写、JSON 文件读写、Word 文件读写、XML 文件读取、CSV 文件读取写入数据库中、数据库数据读取写入 CSV 文件。

在数字校园行走,时而需要走过羊肠小道,时而需要走过天桥,时而需要乘坐电梯,时而需要走楼梯,时而需要走过水上小木桥,不管走哪种道路,同学们总能找到比较适合自己行走的路线。

在 Python 的文件处理中,经常会遇到各种格式的文件处理,如 TXT、CSV、JSON、Word 等,也需要学会找到合适的处理方式。

8.1 with 语句

在开始文件读写操作前,先了解 with 语句的工作原理。

有些任务,可能事先需要设置,事后做清理工作。对于这种场景,Python 的 with 语句提供了一种非常方便的处理方式。一个很好的例子是文件处理获取一个文件句柄,从文件中读取数据,然后关闭文件句柄。如以下示例代码:

```
file = open("/tmp/foo.txt")
data = file.read()
file.close()
```

这里有两个问题:① 可能忘记关闭文件句柄;② 文件读取数据发生异常,没有进行任何处理。对这两个问题做如下异常处理:

```
try:
    f = open('xxx')
except:
    print('fail to open')
    exit(-1)
try:
    # do something
    pass
except:
    # do something
    pass
finally:
    f.close()
```

这段代码可以良好地运行,但是比较冗长。

这时用 with 语句可以更优雅地来处理，还可以很好地处理上下文环境产生的异常。代码更改如下：

```
with open("/tmp/foo.txt") as file:
    data = file.read()
```

代码段变得更加优雅简洁了（具体见 with_theroy_0.py 文件）。

with 工作原理如下：

① 紧跟 with 后面的语句被求值后，返回对象的 __enter__()方法被调用，其返回值将被赋值给 as 后面的变量。

② 当 with 后面的代码块全部被执行完，将调用前面返回对象的 __exit__()方法。

通过如下代码进行理解（with_theory_1.py）：

```
class Sample:
    def __enter__(self):
        print("In __enter__()")
        return "Foo"

    def __exit__(self, type, value, trace):
        print("In __exit__()")

def get_sample():
    return Sample()

with get_sample() as sample:
    print("sample:", sample)
```

执行代码，得到结果如下：

```
In __enter__()
sample: Foo
    In __exit__()
```

正如我们看到的：

① __enter__()方法被执行。

② __enter__()方法返回的值是"Foo"，赋值给变量"sample"。

③ 执行代码，打印变量"sample"的值为"Foo"。

④ __exit__()方法调用 with 的真正强大之处是它可以处理异常。

示例 with_theory_1.py 文件中 Sample 类的 __exit__()方法有 3 个参数：val、type 和 trace。这些参数在异常处理中相当有用。对该示例的代码稍作修改，可以进一步了解 with 是如何工作的，此处不展开讲解，可以查看相关 PY 文件（with_theory_2.py）。

8.2 TXT 文件读写

TXT 文件又称为纯文本文件，一般指以 txt 为后缀的文件。TXT 文件的读写操作比较简单，直接看示例代码即可明白。

为便于本章内容的讲解，这里提前准备了一些基础数据，作为本章各节做文本处理展示操作结果使用，基础数据文件路径为 chapter8/file_read/files。从 Github 把代码 clone

后，在该路径下存放提供好的基础数据文件，同时各种文件读写的源码会存放于 chapter8/file_read 文件夹下。

先看读取 TXT 文件的示例代码（read_txt.py）：

```python
import os

# 取得文件完整路径
txt_file_path = os.path.join(os.getcwd(), 'files/basic_info.txt')


# 定义一个函数
def read_txt_file():
    # 检查 TXT 文件是否存在
    if os.path.exists(txt_file_path) is False:
        return

    # 以读取方式打开 TXT 文件
    with open(txt_file_path, 'r') as r_read:
        # 遍历读取文本内容
        for row in r_read:
            # 打印读取的原始行
            print('分割前数据: {}'.format(row))
            # 对原始行根据空格进行分割
            f_list = row.split(' ')
            # 打印分割的结果列表
            print('根据空格进行分割所得结果为: {}'.format(f_list))
            # 对原始行根据制表符"\t"分割
            field_list = row.split("\t")
            print('根据制表符进行分割所得结果为: {}'.format(field_list))
            # 对原始行"用空白替换,对原始行换行符"\n"用空白替换
            row = row.replace('"', '').replace('\n', '')
            # 替换后的行根据制表符"\t"分割
            replace_field_list = row.split('\t')
            print('替换后分割结果: {}'.format(replace_field_list))
            print('列表长度: {}'.format(len(replace_field_list)))
            full_path_id_str = replace_field_list[2]
            print('数字字符串: {}'.format(full_path_id_str))
            len_num_str = len(full_path_id_str)
            print('数字字符串长度: {}'.format(len_num_str))
            num_str_1_list = full_path_id_str.split('|')
            print('数字字符串分割结果: {}'.format(num_str_1_list))
            # 对数字字符串截取,从第一位截取到倒数第二位
            full_path_id_str = full_path_id_str[1: len_num_str - 1]
            print('截取后数字字符串: {}'.format(full_path_id_str))
            num_str_2_list = full_path_id_str.split('|')
            print('截取后数字字符串分割结果: {}'.format(num_str_2_list))
            # 创建一个 list 对象
            num_list = list()
            for str_i in num_str_2_list:
                num_i = int(str_i)
                num_list.append(num_i)
            print('转换结果: {}'.format(num_list))
```

```
if __name__ == "__main__":
    read_txt_file()
```

该示例代码比较简单,此处不展示执行结果。需要补充说明的一点是,代码中使用了 os.getcwd(),意为获取当前工作目录路径。

```
txt_file_path = os.path.join(os.getcwd(), 'files/basic_info.txt')
```

TXT 文件的写入也比较简单,以下示例将先从给定的 TXT 文件中读取数据,对指定数据适当处理后,将处理结果写入指定 TXT 文件中,write()方法向一个文件中写入数据。示例代码如下(write_txt.py):

```python
import os

txt_file_path = os.path.join(os.getcwd(), 'files/basic_info.txt')
write_txt_file_path = os.path.join(os.getcwd(), 'files/write_txt_file.txt')

# 定义一个函数
def write_txt_file():
    # 检查 TXT 文件是否存在
    if os.path.exists(txt_file_path) is False:
        return

    with open(txt_file_path, 'r') as r_read:
        for row in r_read:
            # 打印读取的原始行
            print(row)
            # 对原始行根据空格进行分割
            f_list = row.split(' ')
            # 打印分割的结果列表
            print('根据空格进行分割所得结果为: {}'.format(f_list))
            # 对原始行根据制表符"\t"分割
            field_list = row.split("\t")
            print('根据制表符进行分割所得结果为: {}'.format(field_list))
            # 对原始行"用空白替换,对原始行换行符"\n"用空白替换
            row = row.replace('"', '').replace('\n', '')
            # 替换后的行根据制表符"\t"分割
            replace_field_list = row.split('\t')
            print('替换后分割结果: {}'.format(replace_field_list))
            print('列表长度: {}'.format(len(replace_field_list)))
            full_path_id_str = replace_field_list[2]
            print('数字字符串: {}'.format(full_path_id_str))
            len_num_str = len(full_path_id_str)
            print('数字字符串长度: {}'.format(len_num_str))
            num_str_1_list = full_path_id_str.split('|')
            print('数字字符串分割结果: {}'.format(num_str_1_list))
            # 对数字字符串截取,从第一位截取到倒数第二位
            full_path_id_str = full_path_id_str[1: len_num_str - 1]
            print('截取后数字字符串: {}'.format(full_path_id_str))
            num_str_2_list = full_path_id_str.split('|')
            print('截取后数字字符串分割结果: {}'.format(num_str_2_list))
            # 直接做转换,代码量少,结果不容易一眼看出
```

```python
            simple_num_list = [int(s) for s in num_str_2_list]
            simple_num_str_list = [s for s in num_str_2_list]
            print('转换结果: {}'.format(simple_num_list))

            # mode='w', 写方式, mode='a', 追加方式打开 JSON 文件
            with open(write_txt_file_path, mode='a') as w_file:
                # 写入数据
                w_file.write(','.join(simple_num_str_list))
                # 换行
                w_file.write('\n')
                print('write sucess.')

    if __name__ == "__main__":
        write_txt_file()
```

执行该示例代码，执行成功后，会在 chapter8/file_read/files 文件夹下生成一个名为 write_txt_file.txt 的文件，其中写入了对应格式的数据。

TXT 是包含极少格式信息的文件，并没有明确的定义，通常是指那些能够被系统终端或者简单的文本编辑器接受的格式。任何能读取文本的程序都能读取 TXT 文件，因此通常认为这种文件是通用的、跨平台的。

由于结构简单，TXT 文件被广泛用于记录信息。在实际生产应用中，更多的是向 TXT 文件中写入数据，一般日志文件都以 TXT 或 LOG 文件为多，能够避免其他文件读写时遇到的一些问题。此外，当文本文件中的部分信息出现错误时，往往能够容易地从错误中恢复，并继续处理其余内容。

8.3　CSV 文件读写

逗号分隔值（Comma-Separated Value，CSV），有时也称为字符分隔值，因为分隔字符也可以不是逗号。CSV 文件以纯文本形式存储表格数据（数字和文本）。

CSV 文件由任意数目的记录组成，记录间以某种换行符分隔。

每条记录由字段组成，字段间的分隔符是其他字符或字符串，最常见的是逗号或制表符。

所有记录都有完全相同的字段序列，通常是纯文本文件。

CSV 文件的规则如下：

① 开头不留空，以行为单位。
② 可含或不含列名，含列名则居文件第一行。
③ 一行数据不跨行，无空行。
④ 以半角逗号（,）作分隔符，列为空也要表达其存在。
⑤ 列内容如存在半角引号（'），则替换成半角双引号（""）转义，即用半角双引号（""）将该字段值包含。
⑥ 文件读写时，引号、逗号操作规则互逆。
⑦ 内码格式不限，可为 ASCII、Unicode 或者其他。
⑧ 不支持特殊字符。

读取 CSV 文件时，开头不留空，以行为单位；可含或不含列名，含列名放第一行。
读取 CSV 文件需要导入 csv 模块，示例代码如下（read_csv.py）：

```python
import os
import csv
import datetime

# 取得文件完整路径
csv_file_path = os.path.join(os.getcwd(), 'files/basic_info.csv')


# 读取 csv 文件
def read_csv_file():
    # 判断对应路径下文件是否存在
    if os.path.exists(csv_file_path) is False:
        return

    # 以读取方式打开 CSV 文件
    with open(csv_file_path, 'r') as r_read:
        # 读取 CSV 文件所有内容
        file_read = csv.reader(r_read)
        # 按行遍历读取内容
        for row in file_read:
            # 查看每行类型及每行长度
            print('CSV 文件读取一行的类型为: {}, 读取一行长度: {}'.format(type(row), len(row)))
            print('CSV 文件读取一行的内容: {}'.format(row))
            # 取得一行中的第三列元素
            full_path_id_str = row[2]
            print(full_path_id_str)
            print('数字字符串: {}'.format(full_path_id_str))
            # 字符串长度
            len_num_str = len(full_path_id_str)
            print('数字字符串长度: {}'.format(len_num_str))
            # 字符串分割
            num_str_1_list = full_path_id_str.split('|')
            print('数字字符串分割结果: {}'.format(num_str_1_list))
            # 对数字字符串截取，从第一位截取到倒数第二位
            num_str = full_path_id_str[1: len_num_str - 1]
            print('截取后数字字符串: {}'.format(num_str))
            num_str_2_list = num_str.split('|')
            print('截取后数字字符串分割结果: {}'.format(num_str_2_list))
            # 直接做转换，代码量少，结果不容易一眼看出
            simple_num_list = [int(s) for s in num_str_2_list]
            print('代码量少的转换结果: {}'.format(simple_num_list))

            # 创建一个 list 对象
            num_list = list()
            for str_i in num_str_2_list:
                num_i = int(str_i)
                num_list.append(num_i)
            print('代码量多，但代码比较清晰易读，转换结果: {}'.format(num_list))
```

```python
        #
        ## 取得读取文件中的时间
        create_time_str = row[7]
        ## 打印字符串的值，并打印字符串类型
        print(create_time_str, type(create_time_str))
        ## 对字符串做类型及格式转换
        create_time = datetime.datetime.strptime(create_time_str, "%Y /%m/%d %H:%M:%S")
        print(create_time, type(create_time))

if __name__ == "__main__":
    read_csv_file()
```

执行该 PY 文件，即可打印从指定文件中读取的数据。

CSV 文件的写入也需要导入 csv 模块，以下示例（write_csv.py）将先从给定的 CSV 文件中读取数据，对指定数据做适当处理后，将处理结果写入指定 CSV 文件中。

```python
import csv
import os

csv_file_path = os.path.join(os.getcwd(), 'files/basic_info.csv')
write_csv_file_path = os.path.join(os.getcwd(), 'files/csv_write.csv')

# 读取 CSV 文件
def write_csv_file():
    # 打开文件并读取内容
    with open(csv_file_path, 'r') as r_read:
        # 读取 CSV 文件所有内容
        file_read = csv.reader(r_read)
        # 按行遍历读取内容
        for row in file_read:
            # 查看每行类型及每行长度
            print('CSV 文件读取一行的类型为：{}，读取一行长度：{}'.format(type(row), len(row)))
            print('CSV 文件读取一行的内容：{}'.format(row))
            # 取得一行中的第三列元素
            full_path_id_str = row[2]
            print(full_path_id_str)
            print('数字字符串：{}'.format(full_path_id_str))
            # 字符串长度
            len_num_str = len(full_path_id_str)
            print('数字字符串长度：{}'.format(len_num_str))
            # 字符串分割
            num_str_1_list = full_path_id_str.split('|')
            print('数字字符串分割结果：{}'.format(num_str_1_list))
            # 对数字字符串截取，从第一位截取到倒数第二位
            num_str = full_path_id_str[1: len_num_str - 1]
            print('截取后数字字符串：{}'.format(num_str))
            num_str_2_list = num_str.split('|')
            print('截取后数字字符串分割结果：{}'.format(num_str_2_list))
            # 直接做转换，代码量少，结果不容易一眼看出
            simple_num_list = [int(s) for s in num_str_2_list]
            simple_num_str_list = [s for s in num_str_2_list]
            print('代码量少的转换结果：{}'.format(simple_num_list))
```

```python
# 创建一个list对象
num_list = list()
for str_i in num_str_2_list:
    num_i = int(str_i)
    num_list.append(num_i)
print('代码量多,但代码比较清晰易读,转换结果: {}'.format(num_list))
csv_data_list = list()
csv_data_list.append(simple_num_str_list)

print(simple_num_str_list)
# mode='w',写方式,mode='a',追加方式打开 CSV 文件,newline='',去除空行
with open(write_csv_file_path, mode='a', newline='') as w_file:
    writer = csv.writer(w_file, dialect='excel')
    for row_item in csv_data_list:
        print(row_item)
        writer.writerow(row_item)

        break

if __name__ == "__main__":
    write_csv_file()
```

执行该示例代码,执行成功后,会在 chapter8/file_read/files 文件夹下生成一个名为 csv_write.csv 的文件,其中写入了对应格式的数据。

在实际应用中,CSV 文件的读写都比较多,特别在数据清洗过程中会更多地应用 CSV 文件来做数据的载体。

比如,从 Hive 导出数据一般选择将 Hive 中的数据批量导入 CSV 文件,再通过读取 CSV 文件批量导入数据库。因为通过一个中间文件作为载体,可以快速地向数据库中插入大批量数据。比如,将 5GB 的数据插入 MySQL,若直接从 Hive 查询数据导入 MySQL,要 5 小时以上,而通过中间文件,从 Hive 将数据写入 CSV 文件耗时为半小时左右,再从 CSV 文件将数据批量导入 MySQL,耗时也仅为半小时左右,整个过程耗时 1 小时左右,速度提升 80%以上。

8.4 JSON 文件读写

JSON 是 JavaScript 的子集,专门用于指定结构化的数据。JSON 是轻量级的数据交换方式,以人们更易读的方式传输结构化数据。

JSON 对象非常像 Python 的字典。

JSON 文件的读操作需要导入 json 模块。示例如下(read_json.py):

```python
import os
import json

# 取得文件完整路径
json_file_path = os.path.join(os.getcwd(), 'files/basic_info.json')
```

```python
def read_json_file():
    if os.path.exists(json_file_path) is False:
        return

    # 以读取方式打开 JSON 文件
    with open(json_file_path, 'r') as r_read:
        # 从 JSON 文件中读取内容,并用 json 模块中的 load 函数转换
        read_result_dict = json.load(r_read)
        # 打印读取 load 所得文本的长度及类型
        print(len(read_result_dict), type(read_result_dict))
        # 取得对应键值
        content_list = read_result_dict.get('RECORDS')
        print(len(content_list), type(content_list))
        # 循环
        for item_dict in content_list:
            print(len(item_dict), type(item_dict))
            print(item_dict)
            full_path_id_str = item_dict.get('full_path_id')
            print(full_path_id_str)
            print('数字字符串: {}'.format(full_path_id_str))
            len_num_str = len(full_path_id_str)
            print('数字字符串长度: {}'.format(len_num_str))
            num_str_1_list = full_path_id_str.split('|')
            print('数字字符串分割结果: {}'.format(num_str_1_list))
            # 对数字字符串截取,从第一位截取到倒数第二位
            num_str = full_path_id_str[1: len_num_str - 1]
            print('截取后数字字符串: {}'.format(num_str))
            num_str_2_list = num_str.split('|')
            print('截取后数字字符串分割结果: {}'.format(num_str_2_list))
            # 直接做转换,代码量少,结果不容易一眼看出
            simple_num_list = [int(s) for s in num_str_2_list]
            print('代码量少的转换结果: {}'.format(simple_num_list))

            # 创建一个 list 对象
            num_list = list()
            for str_i in num_str_2_list:
                num_i = int(str_i)
                num_list.append(num_i)
            print('代码量多,但代码比较清晰易读,转换结果: {}'.format(num_list))

if __name__ == "__main__":
    read_json_file()
```

执行该 PY 文件,即可打印从指定文件中读取的数据。

JSON 文件的写操作也需要导入 json 模块,以下示例(write_json.py)将先从给定的 JSON 文件中读取数据,对指定数据做适当处理后,将处理结果写入指定 JSON 文件中。

```python
import os
import json

json_file_path = os.path.join(os.getcwd(), 'files/basic_info.json')
write_json_file_path = os.path.join(os.getcwd(), 'files/write_json_file.json')
```

```python
def read_json_file():
    if os.path.exists(json_file_path) is False:
        return

    with open(json_file_path, 'r') as r_read:
        # 从 JSON 文件中读取内容,并用 json 模块中的 load 函数转换
        read_result_dict = json.load(r_read)

        # mode='w', 写方式, mode='a', 追加方式打开 JSON 文件
        with open(write_json_file_path, mode='a') as w_file:
            # 通过 json 中的 dumps 函数将数据转换为 JSON 格式,写入 JSON 文件
            w_file.write(json.dumps(read_result_dict))

if __name__ == "__main__":
    read_json_file()
```

执行该示例代码,执行成功后,会在 chapter8/file_read/files 文件夹下生成一个名为 write_json_file.json 的文件,其中写入了对应格式的数据。

8.5 Word 文件读写

DOC 或 DOCX 文件是我们常见的 Word 文件,一般以文章、新闻报道和小说这类文字内容较长的数据为主。

Python 读写 Word 文件需要第三方库扩展支持,需要做如下安装:

```
pip install python-docx
```

读取 Word 文件一般需要如下几个步骤:
(1)生成 Word 对象,并指向 Word 文件。
(2)使用 paragraphs() 获取 Word 对象全部内容。
(3)循环 paragraph 对象,获取每行数据并写入列表。
(4)将列表转换为字符串,每个列表元素使用换行符连接。转换后,数据的段落布局与 Word 文件相似。

读取 Word 文件需要导入 docx 模块。操作示例如下(read_word.py):

```python
import docx
import os

# 取得文件完整路径
file_path = os.path.join(os.getcwd(), 'files/basic_info.doc')

def read_word_file():
    doc = docx.Document(file_path)
    # 遍历所有表格
    for table in doc.tables:
        # 遍历表格的所有行
        for row in table.rows:
            # 一行数据
            row_str = '\t'.join([cell.text for cell in row.cells])
```

```
        print(type(row.cells), len(row.cells))
        print(row.cells[2].text)
        print(row_str)

        full_path_id_str = row.cells[2].text
        print('数字字符串: {}'.format(full_path_id_str))
        len_num_str = len(full_path_id_str)
        print('数字字符串长度: {}'.format(len_num_str))
        num_str_1_list = full_path_id_str.split('|')
        print('数字字符串分割结果: {}'.format(num_str_1_list))
        # 对数字字符串截取,从第一位截取到倒数第二位
        full_path_id_str = full_path_id_str[1: len_num_str - 1]
        print('截取后数字字符串: {}'.format(full_path_id_str))
        num_str_2_list = full_path_id_str.split('|')
        print('截取后数字字符串分割结果: {}'.format(num_str_2_list))
        # 直接做转换,代码量少,结果不容易一眼看出
        simple_num_list = [int(s) for s in num_str_2_list]
        print('代码量少的转换结果: {}'.format(simple_num_list))

        # 创建一个 list 对象
        num_list = list()
        for str_i in num_str_2_list:
            num_i = int(str_i)
            num_list.append(num_i)
        print('代码量多,但代码比较清晰易读,转换结果: {}'.format(num_list))

if __name__ == "__main__":
    read_word_file()
```

执行该 PY 文件,即可打印从指定文件中读取的数据。

Word 文件写入一般需要如下几个步骤:

(1)创建生成临时 Word 对象。

(2)分别使用 add_paragraph()和 add_heading()对 Word 对象添加标题和正文内容。

(3)如果设置正文内容的字体加粗和斜体等,那么可以将正文内容对象的属性 runs[0].bold 和 add_run('XX').italic 设置为 True。

(4)如果插入图片和添加表格,那么可以在 Word 对象中使用 add_picture()和 add_table()方法。

(5)完成数据写入,需要将 Word 对象保存为 Word 文件。

Word 文件写入也需要导入 docx 模块。操作示例如下(write_word.py):

```
import os
from docx import Document
from docx.shared import Inches


def main():
    # 创建文档对象
    document = Document()

    # 设置文档标题,中文要用 Unicode 字符串
    document.add_heading(u'我的一个新文档', 0)
```

```python
    # 往文档中添加段落
    p = document.add_paragraph('This is a paragraph having some ')
    p.add_run('bold ').bold = True
    p.add_run('and some ')
    p.add_run('italic.').italic = True

    # 添加一级标题
    document.add_heading(u'一级标题, level = 1', level=1)
    document.add_paragraph('Intense quote', style='IntenseQuote')

    # 添加无序列表
    document.add_paragraph('first item in unordered list', style='ListBullet')

    # 添加有序列表
    document.add_paragraph('first item in ordered list', style='ListNumber')
    document.add_paragraph('second item in ordered list', style='ListNumber')
    document.add_paragraph('third item in ordered list', style='ListNumber')

    # 添加图片,并指定宽度
    document.add_picture(os.path.join(os.getcwd(),'files/1.jpg'), width= Inches(1.25))

    # 添加表格: 1行3列
    table = document.add_table(rows=1, cols=3)
    # 获取第一行的单元格列表对象
    hdr_cells = table.rows[0].cells
    # 为每一个单元格赋值。注: 值都要为字符串类型
    hdr_cells[0].text = 'Name'
    hdr_cells[1].text = 'Age'
    hdr_cells[2].text = 'Tel'
    # 为表格添加一行
    new_cells = table.add_row().cells
    new_cells[0].text = 'Tom'
    new_cells[1].text = '19'
    new_cells[2].text = '12345678'

    # 添加分页符
    document.add_page_break()

    # 往新的一页中添加段落
    p = document.add_paragraph('This is a paragraph in new page.')

    # 保存文档
    document.save(os.path.join(os.getcwd(), 'files/demo.docx'))

if __name__ == '__main__':
    main()
```

执行该示例代码,执行成功后,会在 chapter8/file_read/files 文件夹下生成一个名为 demo.docx 的文件,其中写入了对应格式的数据。

8.6 XML 文件读取

可扩展标记语言（eXtensible Markup Language，XML）是一个比较老的结构化数据格式，是"纯文本"格式，用来表示结构化的数据。

尽管 XML 是纯文本的，但如果没有解析器的帮助，几乎难以辨认。

XML 文件读取需导入 xml.dom.minidom 模块。操作示例如下（read_xml.py）：

```python
import os
import xml.dom.minidom

# 取得文件完整路径
file_path = os.path.join(os.getcwd(), 'files/basic_info.xml')


def read_xml_file():
    # 使用 minidom 解析器打开 XML 文档
    DOMTree = xml.dom.minidom.parse(file_path)
    collection = DOMTree.documentElement
    print(collection)
    if collection.hasAttribute("id"):
        print("Root element : %s" % collection.getAttribute("id"))

    # 获取集合中所有记录
    record = collection.getElementsByTagName("RECORD")
    # 打印每条详细信息
    for item in record:
        print("value:{}, type:{}".format(item.getElementsByTagName('id'), type(item.getElementsByTagName ('id'))))
        print("value:{}, type:{}".format(item.getElementsByTagName('id')[0], type(item.getElementsByTagName('id')[0])))
        print("value:{}, type:{}".format(item.getElementsByTagName('id') [0].childNodes, type(item.getElementsByTagName('id')[0].childNodes)))
        print("value:{}, type:{}".format(item.getElementsByTagName('id') [0].childNodes[0], type(item.getElementsByTagName('id')[0].childNodes[0])))
        print("value:{}, type:{}".format(item.getElementsByTagName('id') [0].childNodes[0].data, type(item.getElementsByTagName('id')[0].childNodes[0]. data)))

        # getElementsByTagName()方法返回带有指定名称的所有元素的 NodeList
        # childNodes 返回文档的子节点的节点列表
        key_id = item.getElementsByTagName('id')[0].childNodes[0].data
        print("id: {}".format(key_id))
        product_code = item.getElementsByTagName('product_code')[0].childNodes [0].data
        print("product_code: {}".format(product_code))
        full_path_id= item.getElementsByTagName('full_path_id')[0]. childNodes [0].data
        print("full_path_id: {}".format(full_path_id))
        en_name = item.getElementsByTagName('en_name')[0].childNodes[0].data
        print("en_name: {}".format(en_name))
        en_full_path_name = item.getElementsByTagName('en_full_path_name')[0].childNodes[0].data
        print("en_full_path_name: {}".format(en_full_path_name))
        local_file_path = item.getElementsByTagName('local_file_path')[0].childNodes[0].data
```

```python
        print("local_file_path: {}".format(local_file_path))
        modify_time_stamp = item.getElementsByTagName('modify_time_stamp')[0].childNodes[0].data
        print("modify_time_stamp: {}".format(modify_time_stamp))
        create_date = item.getElementsByTagName('create_date')[0].childNodes[0].data
        print("create_date: {}".format(create_date))

        full_path_id_str = full_path_id
        print('数字字符串: {}'.format(full_path_id_str))
        len_num_str = len(full_path_id_str)
        print('数字字符串长度: {}'.format(len_num_str))
        num_str_1_list = full_path_id_str.split('|')
        print('数字字符串分割结果: {}'.format(num_str_1_list))
        # 对数字字符串截取，从第一位截取到倒数第二位
        num_str = full_path_id_str[1: len_num_str - 1]
        print('截取后数字字符串: {}'.format(num_str))
        num_str_2_list = num_str.split('|')
        print('截取后数字字符串分割结果: {}'.format(num_str_2_list))
        # 直接做转换，代码量少，结果不容易一眼看出
        simple_num_list = [int(s) for s in num_str_2_list]
        print('代码量少的转换结果: {}'.format(simple_num_list))

        # 创建一个 list 对象
        num_list = list()
        for str_i in num_str_2_list:
            num_i = int(str_i)
            num_list.append(num_i)
        print('代码量多，但代码比较清晰易读，转换结果: {}'.format(num_list))
        break

if __name__ == "__main__":
    read_xml_file()
```

执行该 PY 文件，即可打印从指定文件中读取的数据。

在实际应用中，通过 Python 代码读取 XML 文件较多，写 XML 的操作非常少，此处不做介绍，有兴趣的读者可以自己查阅相关资料进行研究。

8.7 CSV 文件读取后插入 MySQL 数据库

前面讲解了文本文件的数据读取及数据写入，本节将介绍如何将从 CSV 文件读取的数据插入 MySQL 数据库。

使用 SQLAlchemy 操作数据库，在 chapter8/database 文件夹下存放数据库表操作对象及数据库连接文件。

数据库表创建文件示例代码如下（model_create.py）：

```python
from sqlalchemy import create_engine, Column, String, Integer, func, DateTime, BIGINT
from sqlalchemy.orm import sessionmaker
from sqlalchemy.ext.declarative import declarative_base


def get_db_conn_info():
```

```python
    # "mysql+pymysql://用户名:密码@IP地址/数据库名?charset=UTF8MB4"
    conn_info_r = "mysql+pymysql://root:root@localhost/data_school? charset= UTF8MB4"
    return conn_info_r

conn_info = get_db_conn_info()
engine = create_engine(conn_info, echo=True)

db_session = sessionmaker(bind=engine)
session = db_session()
BaseModel = declarative_base()

class BasicInfo(BaseModel):
    __tablename__ = "basic_info"
    id = Column(Integer, primary_key=True)
    image_id = Column(Integer, default=0, nullable=True, comment='图片id')
    product_code = Column(String(200), default=None, nullable=True, comment='产品代码')
    full_path_id = Column(String(100), default=None, nullable=True, comment='类目结构')
    en_name = Column(String(100), default=None, nullable=True, comment='英文名')
    full_path_en_name = Column(String(200), default=None, nullable=True, comment='全类目英文名')
    file_path = Column(String(300), default=None, nullable=True, comment='路径')
    modify_timestamp = Column(BIGINT, default=0, nullable=False, comment='时间戳')
    create_date = Column(DateTime, default=func.now(), nullable=False, comment='创建时间')
    update_date = Column(DateTime, nullable=True, comment='更改时间')

BaseModel.metadata.create_all(engine)
```

执行该 PY 文件，在数据库 data_school 中会创建一个名为 basic_info 的表。

数据库表操作对象示例代码如下（model_obj.py）：

```python
from sqlalchemy import Column, String, Integer, func, DateTime, BIGINT
from sqlalchemy.ext.declarative import declarative_base

BaseModel = declarative_base()

class BasicInfo(BaseModel):
    __tablename__ = "basic_info"
    id = Column(Integer, primary_key=True)
    image_id = Column(Integer, default=0, nullable=True, comment='图片id')
    product_code = Column(String(200), default=None, nullable=True, comment='产品代码')
    full_path_id = Column(String(100), default=None, nullable=True, comment='类目结构')
    en_name = Column(String(100), default=None, nullable=True, comment='英文名')
    full_path_en_name = Column(String(200), default=None, nullable=True, comment='全类目英文名')
    file_path = Column(String(300), default=None, nullable=True, comment='路径')
    modify_timestamp = Column(BIGINT, default=0, nullable=False, comment='时间戳')
    create_date = Column(DateTime, default=func.now(), nullable=False, comment='创建时间')
    update_date = Column(DateTime, nullable=True, comment='更改时间')
```

数据库连接示例代码如下（sqlalchemy_conn.py）：

```python
from sqlalchemy import create_engine
from sqlalchemy.orm import sessionmaker

# 数据库连接
```

```python
def db_conn():
    conn_info = "mysql+pymysql://root:root@localhost/data_school?charset= UTF8MB4"
    engine = create_engine(conn_info, echo=False)

    db_session = sessionmaker(bind=engine)
    session = db_session()
    return session

# 数据库查询
def query_mysql(sql_str):
    session = db_conn()
    return session.execute(sql_str)

# 数据库更新
def update_mysql(update_sql):
    session = db_conn()
    session.execute(update_sql)
    session.commit()
    session.close()

if __name__ == "__main__":
    print('test')
```

① 按行读取数据插入 MySQL。按行读取，从 CSV 文件每读取一行数据，就往数据库中插入一条记录。示例代码如下（sqlalchemy_csv_insert_mysql.py）：

```python
import time
import csv
import datetime
import os

from chapter8.database.sqlalchemy_conn import db_conn
from chapter8.database.model_obj import BasicInfo

csv_file_path = os.path.join(os.getcwd(), 'files/query_hive.csv')

# 读取 CSV 文件
def read_csv_file():
    start_time = time.time()
    # 打开文件并读取内容
    with open(csv_file_path, 'r') as r_read:
        # 读取 CSV 文件所有内容
        file_read = csv.reader(r_read)
        # 按行遍历读取内容
        row_count = 0
        # 按行读取 CSV 文件内容，并按行插入 MySQL
        for row in file_read:
            if row_count == 0:
                row_count += 1
                print(row)
                continue
```

```python
            row_count += 1
            image_id = row[0]
            file_path = row[1]
            modify_timestamp = row[2]
            product_code = row[3]
            en_name = row[4]
            full_path_id = row[5]
            full_path_en_name = row[6]
            try:
                session = db_conn()
                # 构造插入数据库的语句
                basic_info_obj=BasicInfo(image_id=image_id, file_path=file_path,
                                    modify_timestamp=modify_timestamp,
                                    product_code=product_code,
                                    en_name=en_name, full_path_id=full_path_id,
                                    full_path_en_name=full_path_en_name,
                                    create_date=datetime.datetime.now())
                # 数据按行插入数据库
                session.add(basic_info_obj)
                session.commit()
                session.close()
            except Exception as ex:
                print('insert error:{}'.format(ex))
        print('插入({0})条记录，花费: {1}s'.format(row_count - 1, time.time()- start_time))

if __name__ == "__main__":
    read_csv_file()
```

执行该 PY 文件，得到如下执行结果：

```
['a.imageid', 'a.filepath', 'a.modifytimestamp', 'b.productcode', 'b.enname', 'b.fullpathid', 'b.fullpathenname']
```

插入 2000 条记录，耗时 254.64556503295898s。

由执行结果看到，以按行读取插入数据库的方式，读取并插入 2000 条记录耗时为 254s 左右，即平均每秒不到 10 条。这个耗时与计算机性能有一定关系，但相差不会太大。

② 批量读取数据插入 MySQL：从 CSV 文件中读取一定量的数据后，如 1000 条，以批量方式插入数据库。示例代码如下（sqlalchemy_batch_insert_mysql.py）：

```python
import time
import csv
import datetime
import os

from chapter8.database.sqlalchemy_conn import db_conn
from chapter8.database.model_obj import BasicInfo

csv_file_path = os.path.join(os.getcwd(), 'files/query_hive.csv')

def lines_count():
    """
    CSV 文件总行数统计
    :return: 总行数
```

```python
    """
    f_read = open(csv_file_path, "r")
    cline = 0
    while True:
        buffer = f_read.read(8*1024*1024)
        if not buffer:
            break
        cline += buffer.count('\n')
    f_read.seek(0)
    return cline

# 读取 CSV 文件
def read_csv_file():
    start_time = time.time()
    # CSV 文件总行数统计
    total_line = lines_count()
    # 打开文件并读取内容
    with open(csv_file_path, 'r') as r_read:
        # 读取 CSV 文件全部内容
        file_read = csv.reader(r_read)
        # 按行遍历读取内容
        row_count = 0
        basic_info_obj_list = list()
        for row in file_read:
            if row_count == 0:
                row_count += 1
                print(row)
                continue

            image_id = row[0]
            file_path = row[1]
            modify_timestamp = row[2]
            product_code = row[3]
            en_name = row[4]
            full_path_id = row[5]
            full_path_en_name = row[6]

            basic_info_obj = BasicInfo(image_id=image_id, file_path=file_path,
                                      modify_timestamp=modify_timestamp,
                                      product_code=product_code,en_name=en_name,
                                      full_path_id=full_path_id,
                                      full_path_en_name=full_path_en_name,
                                      create_date=datetime.datetime.now())
            basic_info_obj_list.append(basic_info_obj)
            row_count += 1
            # 每 1000 条记录做一次插入
            if row_count % 1000 == 0:
                batch_insert_into_mysql(basic_info_obj_list)
                basic_info_obj_list.clear()
                continue
```

```python
        # 剩余数据插入数据库
        if row_count == total_line:
            batch_insert_into_mysql(basic_info_obj_list)
            basic_info_obj_list.clear()

    print('插入({0})条记录,花费: {1}s'.format(row_count - 1, time.time()- start_time))

# 数据批量插入数据库
def batch_insert_into_mysql(basic_info_obj_list):
    try:
        session = db_conn()
        session.add_all(basic_info_obj_list)
        session.commit()
        session.close()
    except Exception as ex:
        print('batch insert error:{}'.format(ex))

if __name__ == "__main__":
    read_csv_file()
```

执行该 PY 文件,得到如下执行结果:

```
['a.imageid', 'a.filepath', 'a.modifytimestamp', 'b.productcode', 'b.enname', 'b.fullpathid', 'b.fullpathenname']
插入(2000)条记录,花费: 1.8261044025421143s
```

由执行结果看到,通过批量插入的方式,读取并插入 2000 条记录耗时 1.8s 左右,比按行读取并插入快了 100 多倍。

在实际应用中,遇到大批量数据需要从文本文件导入数据库时,首选批量处理方式,还可以使用多线程或多进程方式加快导入速度。

8.8 CSV 文件读取后插入 MongoDB 数据库

本节将介绍如何将从 CSV 文件读取的数据插入 NoSQL(MongoDB)数据库。

在 chapter8/database 文件夹下存放 NoSQL 数据库连接文件。

MongoDB 数据库连接示例代码如下(mongo_conn.py):

```python
import pymongo

def get_col():
    """
    取得指定集合
    :return: 指定集合
    """
    # mongo client
    mongo_client = pymongo.MongoClient("mongodb://localhost:27017/")
    # 数据库连接,存在 data_school,则直接连接,否则创建
    mongo_conn = mongo_client["data_school"]
    # 集合获取,存在 basic_info 集合,则直接返回,否则创建
    mongo_col = mongo_conn["basic_info"]
    return mongo_col
```

① 按行读取数据插入 MongoDB。同 MySQL 中按行处理一样，读取一条记录插入一条记录。示例代码如下（csv_insert_mongo.py）：

```python
import time
import csv
import datetime
import os

from chapter8.database.mongo_conn import get_col

csv_file_path = os.path.join(os.getcwd(), 'files/query_hive.csv')

# 读取 CSV 文件
def read_csv_file():
    start_time = time.time()
    # 打开文件并读取内容
    with open(csv_file_path, 'r') as r_read:
        # 读取 CSV 文件所有内容
        file_read = csv.reader(r_read)
        # 按行遍历读取内容
        row_count = 0
        # 按行读取 CSV 文件内容，并按行插入 MongoDB
        for row in file_read:
            if row_count == 0:
                row_count += 1
                print(row)
                continue

            row_count += 1
            image_id = row[0]
            file_path = row[1]
            modify_timestamp = row[2]
            product_code = row[3]
            en_name = row[4]
            full_path_id = row[5]
            full_path_en_name = row[6]
            try:
                curr_time = datetime.datetime.now()
                insert_dict = {"image_id": image_id,
                               "file_path": file_path,
                               "modify_timestamp": modify_timestamp,
                               "product_code": product_code,
                               "en_name": en_name,
                               "full_path_id": full_path_id,
                               "full_path_en_name": full_path_en_name,
                               "create_date": curr_time}
                get_col().insert_one(insert_dict)
            except Exception as ex:
                print('insert error:{}'.format(ex))
```

```python
            print('插入({0})条记录,花费: {1}s'.format(row_count - 1, time.time()- start_time))

if __name__ == "__main__":
    read_csv_file()
```

执行该 PY 文件,得到如下执行结果:

```
['a.imageid', 'a.filepath', 'a.modifytimestamp', 'b.productcode', 'b.enname', 'b.fullpathid', 'b.fullpathenname']
插入(2000)条记录,花费: 22.971313953399658s
```

由执行结果可知,MongoDB 中的按行处理速度比 MySQL 快很多,能达到平均每秒约 100 条的速度。

② 批量读取数据插入 MongoDB。同 MySQL 类似,批量读取,再批量插入。示例代码如下(csv_batch_insert_mongo.py):

```python
import time
import csv
import datetime
import os

from chapter8.database.mongo_conn import get_col

csv_file_path = os.path.join(os.getcwd(), 'files/query_hive.csv')

def lines_count():
    """
    CSV 文件总行数统计
    :return: 总行数
    """
    f_read = open(csv_file_path, "r")
    cline = 0
    while True:
        buffer = f_read.read(8*1024*1024)
        if not buffer:
            break
        cline += buffer.count('\n')
    f_read.seek(0)
    return cline

# 读取 CSV 文件
def read_csv_file():
    start_time = time.time()
    # CSV 文件总行数统计
    total_line = lines_count()
    # 打开文件并读取内容
    with open(csv_file_path, 'r') as r_read:
        # 读取 CSV 文件全部内容
        file_read = csv.reader(r_read)
        # 按行遍历读取内容
        row_count = 0
```

```python
            basic_info_obj_list = list()
            # 按行读取 CSV 文件内容，批量插入 MongoDB
            for row in file_read:
                if row_count == 0:
                    row_count += 1
                    print(row)
                    continue

                row_count += 1
                image_id = row[0]
                file_path = row[1]
                modify_timestamp = row[2]
                product_code = row[3]
                en_name = row[4]
                full_path_id = row[5]
                full_path_en_name = row[6]
                try:
                    curr_time = datetime.datetime.now()
                    insert_dict = {"image_id": image_id,
                                   "file_path": file_path,
                                   "modify_timestamp": modify_timestamp,
                                   "product_code": product_code,
                                   "en_name": en_name,
                                   "full_path_id": full_path_id,
                                   "full_path_en_name": full_path_en_name,
                                   "create_date": curr_time}
                    basic_info_obj_list.append(insert_dict)
                    if len(basic_info_obj_list) % 950 == 0:
                        get_col().insert_many(basic_info_obj_list)
                        basic_info_obj_list.clear()
                        continue

                    # 剩余数据插入数据库
                    if row_count == total_line:
                        get_col().insert_many(basic_info_obj_list)
                        basic_info_obj_list.clear()
                except Exception as ex:
                    print('insert error:{}'.format(ex))
            print('插入({0})条记录，花费：{1}s'.format(row_count - 1, time.time()- start_time))

if __name__ == "__main__":
    read_csv_file()
```

执行该 PY 文件，得到如下执行结果：

```
    ['a.imageid', 'a.filepath', 'a.modifytimestamp', 'b.productcode', 'b.enname', 'b.fullpathid',
'b.fullpathenname']
    插入(2000)条记录，花费：0.17701029777526855s
```

由执行结果看到，MongoDB 中的批量插入操作比 MySQL 中批量插入操作快 10 倍以上，非常高效。

8.9 小结

本章主要讲解的是对各种文件中数据的读取和写入，以及文件中数据与关系型数据库（MySQL）及 NoSQL 数据库（MongoDB）的交互。

8.10 实战演练

1. 写代码实现：从 CSV 文件中读取信息，将读取的文本信息写入 TXT 文件。
2. 写代码实现：从数据库中读取信息，写入 CSV 文件。
3. 自己构造一些数据，保存到 CSV 文件，其中至少包含一个数值列和一个可以分组的列。写代码读取 CSV 文件内容，将读取内容进行分组，并求分组中数值列的和，将分组信息和求和结果写入一张数据库表中，数据库表结构由自己设计。

第 9 章 Python 数据爬取

本章将以一个简单的自然语言处理示例展示 Python 中的数据爬取，进一步提高对前面各章知识点的巩固。

本章先介绍几个新的库，如 pyecharts 库、jieba 分词库、BeautifulSoup 库、Requests 库等，接着通过一个爬虫程序爬取相关内容，再将对应内容分词后添加到数据库（关系型及非关系型），最后从数据库中获得数据做词频统计并生成统计图。

数字校园中处处都存在大量数据，在等待同学们去爬取。首先需要同学们知道怎么去爬取，需要什么工具，这些工具怎样使用，从而打造出更好的爬取工具。下面讲解数据爬取工具的打造，以及数据爬取工具的初步试水。

9.1 爬虫基础

在开始分词与词频统计项目之前，我们先了解几个库。

1. pyecharts 库

pyecharts 库是一个用于生成 Echarts 图表的类库。Echarts 是百度开源的一个数据可视化 JS 库。用 Echarts 生成的图可视化效果非常棒，pyecharts 用于与 Python 进行对接，方便在 Python 中直接使用数据生成图。

关于 pyecharts 库的更多信息可以查看官方网站：http://pyecharts.org。

2. jieba 分词库

jieba（结巴）是一个强大的分词库，完美支持中文分词。jieba 支持以下 3 种分词模式。

① 精确模式：将句子以最精确的方式切开，适合文本分析。

② 全模式：把句子中所有可以成词的词语都扫描出来，速度非常快，但是不能解决词语的歧义问题。

③ 搜索引擎模式：在精确模式的基础上，对长词再次切分，提高召回率，适合用于搜索引擎分词。

jieba 分词库还支持繁体分词和自定义词典。

3. BeautifulSoup 库

BeautifulSoup 库是用 Python 写的一个 HTML/XML 解析器，主要功能是从网页爬取需要的数据。

BeautifulSoup 库可以很好地处理不规范标记并生成剖析树（parse tree），提供简单又常用的导航（navigating）、搜索和修改剖析树的操作。

BeautifulSoup 库将 HTML 解析为对象进行处理，全部页面转变为字典或者数组，相对于正则表达式的方式，可以大大简化处理过程。

4. Requests 库

Requests 库是 Python 的一个实用的 HTTP 客户端库，完全满足如今网络爬虫的需求。在开发使用上，Requests 库语法简单易懂，完全符合 Python 优雅、简洁的特性。在兼容性上，Requests 库完全兼容 Python 2 和 Python 3。

9.2 库的安装与使用

本节将讲述如何安装这些库，以及通过简单的示例来介绍这些库的使用方式。

9.2.1 pyecharts 库的安装与使用

在 Python 中，pyecharts 库的安装方式如下：

```
pip install pyecharts
```

pyecharts 库的使用示例如下（pyecharts_exp.py）：

```python
from pyecharts.charts import Bar
from pyecharts import options as opts
from chapter9.config import ROOT_PATH

# 水平横条
bar = Bar()
bar.add_xaxis(["北京","上海","广州","深圳","杭州","成都"])
bar.add_yaxis("阅读人数分布", [500, 450, 360, 450, 355, 380])
bar.set_global_opts(title_opts=opts.TitleOpts(title='Python 3.7 从零开始学', subtitle='按城市统计'))
bar.render(path=ROOT_PATH + '/static/' + "pyecharts_exp.html")
```

在当前目录下创建一个名为 static 的文件夹，执行该示例代码，会在 static 目录下生成一个名为 pyecharts_exp.html 的 HTML 文件。直接用 Chrome 或其他浏览器打开这个 HTML 文件，能看到一张如图 9-1 所示的图。

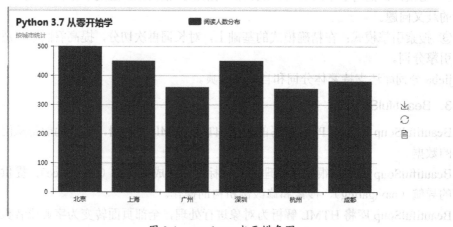

图 9-1　pyecharts 水平横条图

9.2.2 jieba 分词库的安装与使用

在 Python 中，jieba 分词库的安装方式如下：

```
pip install jieba
```

jieba 分词库的使用示例如下（jieba_exp.py）：

```python
import jieba

'''
cut 方法有两个参数
1)第一个参数是我们想分词的字符串
2)第二个参数 cut_all 用来控制是否采用全模式
'''

# 全模式
word_list = jieba.cut(" 《Python 3.7 从零开始学》是一本得好好看看的好书！ ", cut_all=True)
print("全模式: ","|".join(word_list))
#精确模式，默认
word_list = jieba.cut(" 《Python 3.7 从零开始学》是一本得好好看看的好书！ ", cut_all=False)
print("精确模式: ","|".join(word_list))
#搜索引擎模式
word_list = jieba.cut_for_search(" 《Python 3.7 从零开始学》是一本得好好看看的好书！ ")
print("搜索引擎: ","|".join(word_list))
```

执行示例代码，得到如下结果：

```
全模式: |Python|3|7|从零开始|开始|学|||是|一本|得|好好|好好看|好看|看看|的|好书||
精确模式: 《|Python| |3.7| |从零开始|学》|是|一本|得|好好|看看|的|好书|！
搜索引擎: 《|Python| |3.7| |开始|从零开始|学|》|是|一本|得|好好|看看|的|好书|！
```

9.2.3 BeautifulSoup 库的安装与使用

在 Python 中，BeautifulSoup 库的安装方式如下：

```
pip install beautifulsoup4
```

BeautifulSoup 库的使用示例如下（beautiful_soup_exp.py）：

```python
import requests
from bs4 import BeautifulSoup

url = 'http://www.baidu.com'
# 创建实例
resp = requests.get(url)
# 创建对象
bs = BeautifulSoup(resp.content)
# 提取 Tag
title_cont = bs.title
print(title_cont)
print(type(title_cont))
```

执行该示例代码，得到如下结果：

```
<title>百度一下，你就知道</title>
<class 'bs4.element.Tag'>
```

9.2.4 Requests 库的安装与使用

Requests 库可以通过 pip 安装，安装语句如下：

```
pip install requests
```

HTTP 常用请求有 GET 和 POST 两种，Requests 库区分不同的请求方式，下面以 GET 请求为例介绍。

Requests 库实现 GET 请求示例如下（get_exp.py）：

```python
import requests

# 不带参数的请求方式
req = requests.get('https://www.baidu.com')

# 带参数的请求方式
url_1 = 'https://www.baidu.com/s?wd=python'
req_1 = requests.get(url_1)

# 带参数的请求方式
url_2 = 'https://www.baidu.com/s'
params = {'wd': 'python'}
req_2 = requests.get(url_2, params=params)
Requests 实现 POST 请求示例（post_exp.py）：
import requests
import json

# 定义参数字典
data = {'key1': 'value1', 'key2': 'value2'}
# 将字典转换为 JSON
data = json.dumps(data)

url = 'https://www.baidu.com/'
# 发送 post 请求
req = requests.post(url, data=data)
print(req.text)
```

9.3 分词与词频统计实战

经过前面两节知识点的讲解，现在我们结合这些知识点实现以下功能：

① 从一个指定网站爬取对应数据，本章以爬取百度知道的内容为例，根据输入指定关键字（关键字不能为空），爬取用该关键字搜索到的问题标题和问题回答，并对问题标题以搜索引擎模式进行分词。

② 将①中爬取的问题标题、问题回答和分词结果保存到 MySQL 数据库中。

③ 数据爬取结束并保存到数据库后，从 MySQL 数据库中取出保存的记录，根据保存的分词结果进行统计，经过给定的关键词库进行过滤后，统计各关键词出现的次数，并以图表的形式展现出来。

根据以上要求，下面介绍具体的实现方式。

9.3.1 整体结构设计

在 chapter9 文件夹下创建一些文件或文件夹，功能介绍分别如下。

① database 文件夹用于存放与数据库直接关联的文件。models.py 文件用于编写模型对象，即数据库表对应的对象，以及表的增、删、改、查的编写。mongo_conn.py 文件用于编写 MongoDB 数据库连接代码。

② rule 文件夹用于存放定义的规则，key_words.py 文件用于编写关键词库的集合。

③ server 文件夹用于编写操作 MySQL 数据库的逻辑业务。get_input_info.py 文件用于读取从控制台输入的参数值。info_search.py 文件用于编写数据爬取代码，以及将爬取数据处理后保存到数据库的逻辑。word_count.py 文件用于编写词频统计和生成图表逻辑。

④ mongo_server 文件夹用于编写操作 MongoDB 数据库的逻辑业务。get_input_info.py 文件用于读取从控制台输入的参数值。mongo_info_search.py 文件用于编写数据爬取代码，以及将爬取数据处理后保存到数据库的逻辑。mongo_word_count.py 文件用于编写词频统计和生成图表逻辑。

⑤ static 文件夹用于存放静态 HTML 文件。

⑥ mongo_run.py 文件为使用非关系型数据库进行操作的项目入口。

⑦ run.py 文件为使用关系型数据库进行操作的项目入口。

其完整代码将在 9.4 节展示，下面对其逻辑进行讲解，并以代码辅助。

9.3.2 数据结构设计

根据前面的整体结构设计，数据结构设计代码编写在 models.py 文件中，定义了 NLPAnalysis 类，类中定义 __tablename__ 的值为 nlp_analysis，即表名为 nlp_analysis。再定义 4 个字段，命名分别为 id（主键，Integer 类型）、question_title（问题标题，String，长度为 200）、question_answer（问题答案，String，长度为 500）和 fen_ci_result（标题分词结果，String，长度为 1000）。代码实现如下（models.py）：

```
from sqlalchemy import create_engine, Column, String, Integer
from sqlalchemy.orm import sessionmaker
from sqlalchemy.ext.declarative import declarative_base

# 建立链接
engine=create_engine('mysql+pymysql://root:root@localhost/test?charset=utf8',
                    echo=False,pool_size = 5)
# 建立会话
DBSession = sessionmaker(bind=engine)
session = DBSession()
# 模型声明
Base = declarative_base()

class NLPAnalysis(Base):
    __tablename__ = 'nlp_analysis'
```

```python
    id = Column(Integer, primary_key=True)
    question_title = Column(String(200), default=None, doc='问题标题')
    question_answer = Column(String(500), default=None, doc='问题答案')
    fen_ci_result = Column(String(1000), default=None, doc='标题分词结果')

# drop_all 根据模型用来删除表，该语句慎用，此处为示例而用，一般不建议使用
Base.metadata.drop_all(engine)
# 根据模型用来创建表
Base.metadata.create_all(engine)
```

若使用 MongoDB，不需要设计数据结果，但需要获取数据库和集合的连接。封装数据库、集合获取代码示例如下（mongo_conn.py）：

```python
import pymongo

def get_col():
    """
    取得指定集合
    :return: 指定集合
    """
    # mongo client
    mongo_client = pymongo.MongoClient("mongodb://localhost:27017/")
    # 数据库连接，若存在 data_school，则直接连接，否则创建
    mongo_conn = mongo_client["data_school"]
    # 集合获取，若存在 nlp_analysis 集合，则直接返回，否则创建
    mongo_col = mongo_conn["nlp_analysis"]
    return mongo_col
```

9.3.3 数据的爬取与保存

根据从 get_input_info.py 文件中获取的输入参数，去指定网站根据指定关键字进行数据爬取。数据爬取的关键点如下。

分析网页，分析需要用 GET 请求还是 POST 请求，如 https://zhidao.baidu.com/。经分析，要使用的是 GET 请求（此处不做具体爬虫方法的介绍，可自行查询资料解决，如在浏览器搜索：查看 HTTP 请求详情，会显示很多答案）。接着，分析请求头的构造形式，再分析进行关键字搜索时，参数的构造形式是怎样的，关键字怎样放入 URL 请求中，定位到某一页的请求参数是怎样的，由此构造一个通用的字符串参数。最后，分析问题标题和问题答案所对应的是哪些标签的内容。

这部分逻辑代码实现如下，定义 get_data_from_web 方法（info_search.py）：

```python
# 数据收集
def get_data_from_web(input_key_word, begin_page=None, end_page=None):
    for i in range(begin_page, end_page):
        if begin_page is not None and begin_page > 0 and i < begin_page - 1:
            continue

        search_url = (BAIDU_PRE + BAIDU_SEARCH).format(input_key_word, i * 10)
        print(f'当前爬取第({i})页，搜索 URL 为:{search_url}')
        try:
            r = requests.get(search_url, headers=headers)
```

```
            status_code = r.status_code
            if status_code != 200:
                return

            req = BeautifulSoup(r.content.decode('gbk', 'ignore'), 'html5lib')
            result_item_val = req.find_all('div', re.compile('list-inner'))[0]
            result_item_list = result_item_val.find_all('div', re.compile ('list'))[0]
            a_tag_list = result_item_list.find_all('a', re.compile('ti'))
            for a_tag_item in a_tag_list:
                if a_tag_item is None or a_tag_item == '':
                    continue

                href_val = str(a_tag_item.get('href'))
                if href_val is None or href_val == '':
                    continue
                # 问题标题
                question_title = a_tag_item.text
                # 最多取 200 个字符
                if len(question_title) > 200:
                    question_title = question_title[ : 200]
                # 问题答案
                question_answer = get_detail_info(href_val)
                if len(question_answer) > 500:
                    question_answer = question_answer[ : 500]
                # 问题标题分词结果
                fen_ci_result = jie_ba_fen_ci(question_title)
        except Exception as ex:
            print(f'爬取第({i})页失败,失败原因: {ex}')
        print(f'第({i})页信息爬取结束。')
```

其中，get_detail_info()方法实现如下：

```
    detail_url = detail_suffix
    r = requests.get(detail_url, headers=headers)
    resp = BeautifulSoup(r.content.decode('gbk', 'ignore'), 'html5lib')
    detail_text_list = resp.find_all('div', re.compile('best-text'))
    if detail_text_list is None or detail_text_list.__len__() <= 0:
        return ''

    question_answer = str(detail_text_list[0].text).strip()

    return question_answer
```

jie_ba_fen_ci()方法实现如下：

```
# jie ba 分词
def jie_ba_fen_ci(input_val):
    # 搜索引擎模式
    result_list = jieba.cut_for_search(input_val)
    result_val = ','.join(result_list)
    return result_val
```

问题标题、问题答案和分词结果取得后，接下来需要把这些获取的信息保存到数据库中。保存数据时，将需要保存的数据以一个对象的形式传递给模型，在 get_data_from_ web()方法中需要加入以下代码：

```python
row_info_dict = dict()
row_info_dict['question_title'] = question_title
row_info_dict['question_answer'] = question_answer
row_info_dict['fen_ci_result'] = fen_ci_result
models.insert_record(row_info_dict, 'nlp_analysis')
```

同时，需要在 models.py 中添加以下方法：

```python
def insert_record(dict_value, table_name=None):
    """
    插入记录方法
    :param dict_value: 字段值字典
    :param table_name: 表名
    :return: None
    """
    if table_name == 'nlp_analysis' and dict_value:
        data = NLPAnalysis(question_title=dict_value['question_title'],
                          question_answer=dict_value['question_answer'],
                          fen_ci_result=dict_value['fen_ci_result'])
        session.add(data)
        session.commit()
        session.close()
```

至此实现了数据的爬取和数据的持久化，完整代码可在 9.4 节中查看。

若使用 MongoDB 作为数据库，则只需将 get_data_from_web()方法中的代码：

```python
models.insert_record(row_info_dict, 'nlp_analysis')
```

更改为：

```python
get_col().insert_one(row_info_dict)
```

并导入数据库，连接获取模块即可。

9.3.4 制定关键词库

爬取数据后，需要分析哪些关键词对于词频分析是有用的，哪些是无用的，有用的保留，无用的则过滤。下面是一个简单的关键词搜集示例，这些可以视为有用的关键词，代码如下（key_words.py）：

```python
useful_word_list = ['Python', 'python', '3.5', '3.7', '升级', '入门', '基础', '精通', '实现',
                    '为什么', '是什么', '关于', '意思', '问题', '教程', '试学']
```

9.3.5 词频统计与图表生成

有了数据和关键词比对规则后，接下来要做的是从数据库取出数据，根据过滤规则过滤后，统计满足规则的各词的出现次数。定义 word_count()方法，代码如下（word_count.py）：

```python
def word_count():
    word_tuple_list = query_from_mysql()
    for word_tuple in word_tuple_list:
```

```python
        if word_tuple is None or len(word_tuple) <= 0:
            continue
        word_list = word_tuple[0].split(',')
        for item_val in word_list:
            if item_val is None or item_val == '' or str(item_val).strip() == '':
                continue

            # 有用字符匹配
            val_in_list = item_val in key_words.useful_word_list
            if val_in_list is False:
                continue

            count_num = word_dict.get(item_val)
            if count_num is not None and count_num >= 1:
                count_num += 1
                word_dict[item_val] = count_num
            else:
                word_dict[item_val] = 1
```

query_from_mysql()的实现如下：

```python
def query_from_mysql():
    """
    从表中查询结果
    :return: tuple 集列表
    """
    query_sql = "SELECT fen_ci_result FROM nlp_analysis"
    result_list = models.query_record(query_sql)
    return result_list
```

若使用 MongoDB，则 query_from_mysql()的实现示例如下：

```python
def query_from_mysql():
    """
    从表中查询结果
    :return: tuple 集列表
    """
    result_list = get_col().find({}, {"_id": 0, "fen_ci_result": 1})
    return result_list
```

在 models.py 文件中需要添加如下方法：

```python
def query_record(query_sql):
    """
    记录查询方法
    :param query_sql: 查询语句
    :return: 查询结果
    """
    return session.execute(query_sql)
```

统计工作完成后，最后需要根据统计结果绘制图表。以绘制水平图表为例，实现代码如下：

```python
# 水平图表
def draw_bar_horizontal():
    word_items = list(word_dict.items())
```

```python
        word_keys = [k for k, v in word_items]
        word_values = [v for k, v in word_items]
        bar = Bar()
        bar.add_xaxis(word_keys)
        bar.add_yaxis('引用次数', word_values)
        bar.set_global_opts(title_opts=opts.TitleOpts(title='水平图表', subtitle='关键字使用情况分布'))
        # HTML 文件存放路径
        bar.render(path=file_path_pre + "bar_horizontal.html")
```

在 word_count()方法的最后需要调用该方法,以达到统计结束后就绘制统计图的效果。完整代码见 9.4 节。

9.4 分词和词频统计的完整代码

经过 9.3 节的详细讲解,我们对 9.3 节提出的需求已经有了一个大致的实现思路,本节将展现各 PY 文件的完整代码。

1. MySQL 操作处理的完整代码

程序入口代码如下(run.py):

```python
from chapter9.server.get_input_info import get_input_data
from chapter9.server.info_search import get_data_from_web
from chapter9.server.word_count import word_count

if __name__ == "__main__":
    # 取得输入参数
    input_key_word, begin_page, end_page = get_input_data()
    # 从网站取得数据并存储到数据库
    get_data_from_web(input_key_word, begin_page, end_page)
    # 词频统计并生成统计图
    word_count.word_count()
```

根路径配置代码如下(config.py):

```python
import os

# 根路径
curr_path = os.getcwd()
root_index = curr_path.find('chapter9') + len('chapter9')
ROOT_PATH = curr_path[:root_index]
```

模型实现代码如下(models.py):

```python
from sqlalchemy import create_engine, Column, String, Integer
from sqlalchemy.orm import sessionmaker
from sqlalchemy.ext.declarative import declarative_base

# 建立链接
engine=create_engine('mysql+pymysql://root:root@localhost/test?charset=utf8',echo=False,pool_size=5)
# 建立会话
DBSession = sessionmaker(bind=engine)
session = DBSession()
```

```python
# 模型声明
Base = declarative_base()

class NLPAnalysis(Base):
    __tablename__ = 'nlp_analysis'
    id = Column(Integer, primary_key=True)
    question_title = Column(String(200), default=None, doc='问题标题')
    question_answer = Column(String(500), default=None, doc='问题答案')
    fen_ci_result = Column(String(1000), default=None, doc='标题分词结果')

# drop_all 根据模型用来删除表，该语句慎用，此处为示例而用，一般不建议使用
Base.metadata.drop_all(engine)
# 根据模型用来创建表
Base.metadata.create_all(engine)

def insert_record(dict_value, table_name=None):
    """
    插入记录方法
    :param dict_value: 字段值字典
    :param table_name: 表名
    :return: None
    """
    if table_name == 'nlp_analysis' and dict_value:
        data = NLPAnalysis(question_title=dict_value['question_title'],
                           question_answer=dict_value['question_answer'],
                           fen_ci_result=dict_value['fen_ci_result'])
        session.add(data)
        session.commit()
        session.close()

def query_record(query_sql):
    """
    记录查询方法
    :param query_sql: 查询语句
    :return: 查询结果
    """
    return session.execute(query_sql)
```

取得输入参数实现代码如下（get_input_info.py）：

```python
BAIDU_MAX_PAGE_NUM = 70

# 取得输入参数
def get_input_data():
    input_str = input("请输入关键词: ")
    input_str = input_str.strip()
    while input_str is None or input_str == '':
        input_str = input("请输入关键词: ")

    begin_page = input("请输入起始页码(页码必须为大于等于0的数字, "
                       "不输入直接按 Enter，默认为0): ")
    begin_page = begin_page.strip()
```

```python
        if begin_page is None or begin_page == '':
            begin_page = 0
        while number_judge(begin_page) is False:
            begin_page = input("输入不是数字，请输入起始页码(页码必须为大于""等于0的数字，默认为0)：")
            end_page = input("请输入结束页码(结束页码必须大于等于起始页码，最大为{}, "
                             "不输入直接按 Enter，默认为最大值。超过最大值，"
                             "以最大值算)：".format(BAIDU_MAX_PAGE_NUM))
        end_page = end_page.strip()
        if end_page is None or end_page == '':
            end_page = BAIDU_MAX_PAGE_NUM
        while number_judge(end_page) is False:
            end_page = input("输入不是数字，请输入结束页码：")
        if int(end_page) > BAIDU_MAX_PAGE_NUM:
            end_page = BAIDU_MAX_PAGE_NUM

        begin_page = int(begin_page)
        end_page = int(end_page)
        if begin_page < 0:
            begin_page = 0

        if end_page <= begin_page:
            end_page = begin_page + 1

        print(f'起始页码为：{begin_page}，结束页码为：{end_page}')
        return input_str, begin_page, end_page

def number_judge(input_val):
    try:
        float(input_val)
        return True
    except ValueError:
        pass

    try:
        import unicodedata
        unicodedata.numeric(input_val)
        return True
    except (TypeError, ValueError):
        pass

    return False
```

从网站取得数据并存储到数据库的实现代码如下（info_search.py）：

```python
import requests
import re
import jieba
from chapter9.database import models
from bs4 import BeautifulSoup

# 请求头
headers = {
    'Accept-Encoding': 'gzip, deflate, sdch, br',
```

```python
    'Cookie': 'appver=1.5.0.75771',
    'Content-Type': 'text/html',
    'Accept-Language': 'zh-CN,zh;q=0.8',
    'Cache-Control': 'max-age=0',
    'User-Agent': 'Mozilla/5.0 (Windows NT 6.1; WOW64) AppleWebKit/537.36 '
                  '(KHTML, like Gecko) Chrome/57.0.2987.133 Safari/537.36'
}

# URL 前缀
BAIDU_PRE = 'https://zhidao.baidu.com/'
# URL 前缀
BAIDU_SEARCH = 'search?word={}&ie=gbk&site=-1&sites=0&date=0&pn={}'

# 数据收集
def get_data_from_web(input_key_word, begin_page=None, end_page=None):
    for i in range(begin_page, end_page):
        if begin_page is not None and begin_page > 0 and i < begin_page - 1:
            continue

        search_url = (BAIDU_PRE + BAIDU_SEARCH).format(input_key_word, i * 10)
        print(f'当前爬取第({i})页,搜索url为:{search_url}')
        try:
            r = requests.get(search_url, headers=headers)
            status_code = r.status_code
            if status_code != 200:
                return

            req = BeautifulSoup(r.content.decode('gbk', 'ignore'), 'html5lib')
            result_item_val = req.find_all('div', re.compile('list-inner'))[0]
            result_item_list = result_item_val.find_all('div', re.compile ('list')) [0]
            a_tag_list = result_item_list.find_all('a', re.compile('ti'))
            for a_tag_item in a_tag_list:
                if a_tag_item is None or a_tag_item == '':
                    continue

                href_val = str(a_tag_item.get('href'))
                if href_val is None or href_val == '':
                    continue
                # 问题标题
                question_title = a_tag_item.text
                # 最多取 200 个字符
                if len(question_title) > 200:
                    question_title = question_title[ : 200]
                # 问题答案
                question_answer = get_detail_info(href_val)
                if len(question_answer) > 500:
                    question_answer = question_answer[ : 500]
                # 问题标题分词结果
                fen_ci_result = jie_ba_fen_ci(question_title)
                if len(fen_ci_result) > 1000:
                    fen_ci_result = fen_ci_result[ : 1000]
                row_info_dict = dict()
```

```python
                    row_info_dict['question_title'] = question_title
                    row_info_dict['question_answer'] = question_answer
                    row_info_dict['fen_ci_result'] = fen_ci_result
                    models.insert_record(row_info_dict, 'nlp_analysis')
        except Exception as ex:
            print(f'爬取第({i})页失败,失败原因:{ex}')
        print(f'第({i})页信息爬取结束。')

def get_detail_info(detail_suffix):
    detail_url = detail_suffix
    r = requests.get(detail_url, headers=headers)
    resp = BeautifulSoup(r.content.decode('gbk', 'ignore'), 'html5lib')
    detail_text_list = resp.find_all('div', re.compile('best-text'))
    if detail_text_list is None or detail_text_list.__len__() <= 0:
        return ''

    question_answer = str(detail_text_list[0].text).strip()

    return question_answer

# jieba 分词
def jie_ba_fen_ci(input_val):
    # 搜索引擎模式
    result_list = jieba.cut_for_search(input_val)
    result_val = ','.join(result_list)
    return result_val

if __name__ == "__main__":
    # 从网站取得数据并存储到数据库
    get_data_from_web('', 0, 1)
```

有效关键词实现代码如下(**key_words.py**):

```python
useful_word_list = ['Python', 'python', '3.5', '3.7', '升级', '入门', '基础', '精通', '实现',
                    '为什么', '是什么', '关于', '意思', '问题', '教程', '试学']
```

词频统计并生成统计图,为了更直观地查看统计图效果,本例中生成 3 种统计图:水平图表、饼图、词云图。代码实现如下(**word_count.py**):

```python
import os
from chapter9.database import models
from chapter9.rule import key_words
from pyecharts import Bar, Pie, WordCloud
from pyecharts import options as opts
from chapter9.config import ROOT_PATH

file_path_pre = os.path.join(ROOT_PATH, 'static/')

# 关键词统计字典
word_dict = {}

def query_from_mysql():
    """
    从表中查询结果
    :return: tuple 集列表
```

```python
    """
    query_sql = "SELECT fen_ci_result FROM nlp_analysis"
    result_list = models.query_record(query_sql)
    return result_list


def word_count():
    word_tuple_list = query_from_mysql()
    for word_tuple in word_tuple_list:
        if word_tuple is None or len(word_tuple) <= 0:
            continue
        word_list = word_tuple[0].split(',')
        for item_val in word_list:
            if item_val is None or item_val == '' or str(item_val).strip() == '':
                continue

            # 有用字符匹配
            val_in_list = item_val in key_words.useful_word_list
            if val_in_list is False:
                continue
            count_num = word_dict.get(item_val)
            if count_num is not None and count_num >= 1:
                count_num += 1
                word_dict[item_val] = count_num
            else:
                word_dict[item_val] = 1
    draw_bar_horizontal()
    draw_pie()
    draw_word_cloud()

# 水平图表
def draw_bar_horizontal():
    word_items = list(word_dict.items())
    word_keys = [k for k, v in word_items]
    word_values = [v for k, v in word_items]
    bar = Bar()
    bar.add_xaxis(word_keys)
    bar.add_yaxis('引用次数', word_values)
    bar.set_global_opts(title_opts=opts.TitleOpts(title='水平图表', subtitle='关键字使用情况分布'))
    # HTML 文件存放路径
    bar.render(path=file_path_pre + "bar_horizontal.html")

# 饼图
def draw_pie():
    word_items = list(word_dict.items())
    word_keys = [k for k, v in word_items]
    word_values = [v for k, v in word_items]
    data_pairs = [list(z) for z in zip(word_keys, word_values)]
    pie = Pie()
    pie.add("引用次数", data_pairs)
    pie.set_global_opts(title_opts=opts.TitleOpts(title='饼图'))
    # HTML 文件存放路径
    pie.render(path=file_path_pre + "pie.html")
```

```python
# 词云图
def draw_word_cloud():
    word_items = list(word_dict.items())
    word_keys = [k for k, v in word_items]
    word_values = [v for k, v in word_items]
    data_pairs = [list(z) for z in zip(word_keys, word_values)]
    word_cloud = WordCloud()
    word_cloud.add("", data_pairs, word_size_range=[20, 100])
    word_cloud.set_global_opts(title_opts=opts.TitleOpts(title='词云图'))
    # HTML 文件存放路径
    word_cloud.render(path=file_path_pre + "word_cloud.html")

if __name__ == "__main__":
    word_count()
```

执行 run.py 文件，从控制台输入如下信息：

```
请输入关键词: Python
请输入起始页码(页码必须为大于等于 0 的数字，不输入直接按 Enter，默认为 0): 0
请输入结束页码(结束页码必须大于等于起始页码，最大为 70，不输入直接按 Enter，默认为最大值。超过最大值，以最大值算): 2
```

输入信息后，控制台打印出如下信息：

```
起始页码为: 0，结束页码为: 2
当前爬取第(0)页，搜索 url 为:
https://zhidao.baidu.com/search?word=python&ie=gbk&site=-1&sites=0&date=0&pn=0
Building prefix dict from the default dictionary ...
Loading model from cache C:\Users\lyz\AppData\Local\Temp\jieba.cache
Loading model cost 1.237 seconds.
Prefix dict has been built succesfully.
第(0)页信息爬取结束。
当前爬取第(1)页，搜索 URL 为:
https://zhidao.baidu.com/search?word=python&ie=gbk&site=-1&sites=0&date=0&pn=10
第(1)页信息爬取结束。
```

在 static 文件夹下将生成如图 9-2 所示的静态 HTML 文件。

用 Chrom 浏览器打开 bar_horizontal.html 文件，得到如图 9-3 所示的水平图表。

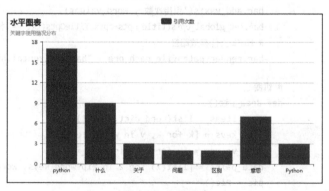

图 9-2 静态 HTML 文件   图 9-3 水平图表

用 Chrom 浏览器打开 pie.html 文件，得到如图 9-4 所示的饼图统计图。

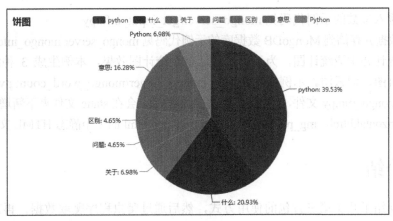

图 9-4 饼图统计图

用 Chrom 浏览器打开 word_cloud.html 文件,得到如图 9-5 所示的词云图统计图。

图 9-5 词云图统计图

至此完成了分词和生成统计图的功能。

在该项目中也可以生成更多其他形式的统计图,也可以将该项目扩展为功能更加强大的项目,可以写入 CSV 文件和发送邮件,加入提醒相关人员对应的信息的功能。

2. MongoDB 操作处理的完整代码实现

程序入口实现示例代码如下(mongo_run.py):

```python
from chapter9.mongo_server.get_input_info import get_input_data
from chapter9.mongo_server.mongo_info_search import get_data_from_web
from chapter9.mongo_server.mongo_word_count import word_count

if __name__ == "__main__":
    # 取得输入参数
    input_key_word, begin_page, end_page = get_input_data()
    # 从网站取得数据并存储到数据库
    get_data_from_web(input_key_word, begin_page, end_page)
    # 词频统计并生成统计图
    word_count()
```

接下来的 PY 文件与关系型数据库中对应 PY 文件的差别不大,此处不具体提供示例代码,只提供示例代码文件名,读者可以从 Github 上 clone。

取得 MongoDB 数据库连接的示例代码见 mongo_conn.py。

取得输入参数的示例代码见 mongo_server/get_input_info.py。

取得数据并存储到 MongoDB 数据库的示例代码见 mongo_server/mongo_info_search.py。

词频统计并生成统计图，为了更直观地查看统计图效果，本例生成 3 种统计图：水平图表、饼图、词云图。示例代码实现见 mongo_server/mongo_word_count.py。

执行 mongo_run.py 文件，输入对应的输入参数后，会在 static 文件夹下新增名称分别为 mg_bar_horizontal.html、mg_pie.html、mg_word_cloud.html 的 3 个静态 HTML 文件。

9.5 小结

本章介绍了几个第三方包的使用方式，然后通过爬虫程序爬取数据，再对数据进行分词处理并保存到数据库，最后对各词做词频统计及生成统计图表。

该项目也可以加入很多可以扩展的功能，有兴趣的读者可以自行加入更多功能，形成一个更强大的项目。

9.6 实战演练

1. 尝试使用 pyecharts 将自己上学期期末考试的各科目与分数通过图表展示出来，图表自选。

2. 挑几句有歧义的语句，使用 jieba 进行分词，查看分词结果，并思考是否可以改进分词结果。

3. 模仿书中示例，从新浪微博或开源中国爬取一些文章，对爬取的文章使用 jieba 分词，做词频统计，并使用 pyecharts 展示各词的词频。

第 10 章　项目实战：音乐数据爬取

经过前面各章的知识点的讲解与练习，本章通过项目实战的方式，对前面所学知识点进一步应用。

本章主要讲解如何从 QQ 音乐网站爬取歌曲、歌手等信息，并将这些爬取的信息在关系型数据库和非关系型数据库中做持久化。

说明：因为本书采用黑白印制，无法体现项目的彩色图片，故采用二维码形式提供；同时，本章及前几章的源代码也以二维码形式提供。

图片

源代码

数字校园里有一个音乐池，其中生活着各种数据鱼。若想要从音乐池中捕获对应种类的数据鱼，需要编织只能捕获该类鱼的爬虫网，这种爬虫网一旦编织而成，可以将音乐池中所有这种类型的鱼都捕获，基本不会有漏网之鱼。接下来开始你的捕鱼之旅。

10.1　获取全部歌手

做网站信息爬取之前，我们先大概了解网页信息的分析方式。

打开 QQ 音乐官网，进入歌手页，网址如下：https://y.qq.com/portal/singer_list.html。在 Chrome 浏览器下打开该网址，按 F12 键，可以看到类似图 10-1 所示的界面。在页面的右侧可以看到一个可选择框，即开发者工具。从开发者工具中可以看到 All、XHR、JS、CSS、Img、Media 等选项。

左下方的 name 中是页面展示所需资源的链接，右下方中有 Headers、Response 等选项，可以查看各链接资源的请求头信息、网页源码信息、响应页面信息等。

如图 10-2 所示，右边用椭圆圈起的 General 部分展示的是歌手列表的 Request URL。

找到歌手列表的 URL 后，单击右侧的 Preview 选项，如图 10-3 所示，可以看到歌手列表信息展示的数据结构形式，singerlist 对应的结构。

由这里可以分析歌手列表的 URL 形式如下：

```
QQ_MUSIC_SINGER_URL = 'https://u.y.qq.com/cgibin/musicu.fcg?callback=getUCGI1986419190708677&g_tk=69824246&jsonpCallback=getUCGI1986419190708677&loginUin=825393011&hostUin=0&format=jsonp&inCharset=utf8&outCharset=utf-8&notice=0&platform=yqq&needNewCode=0&data=%7B%22comm%22%3A%7B%22ct%22%3A%24%2C%22cv%22%3A10000%7D%2C%22singerList%22%3A%7B%22module%22%3A%22Music.SingerListServer%22%2C%22method%22%3A%22get_singer_list%22%2C%22param%22%3A%7B%22area%22%3A-100%2C%22sex%22%3A-100%2C%22genre%22%3A-100%2C%22index%22%3A-100%2C%22sin%22%3A%3A0%2C%22cur_page%22%3A%3A1%7D%7D%7D'
```

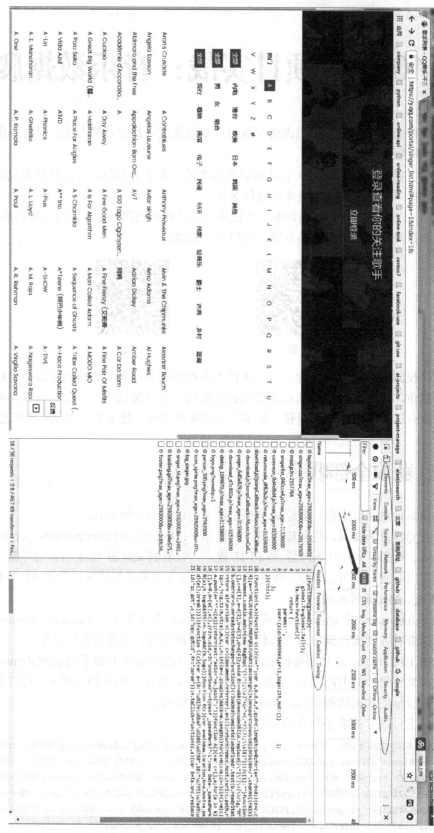

图 10-1 QQ 音乐歌手页

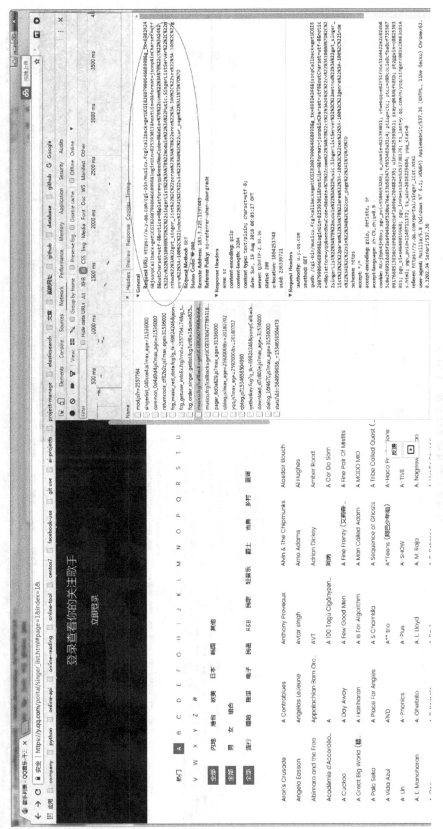

图 10-2 Request URL

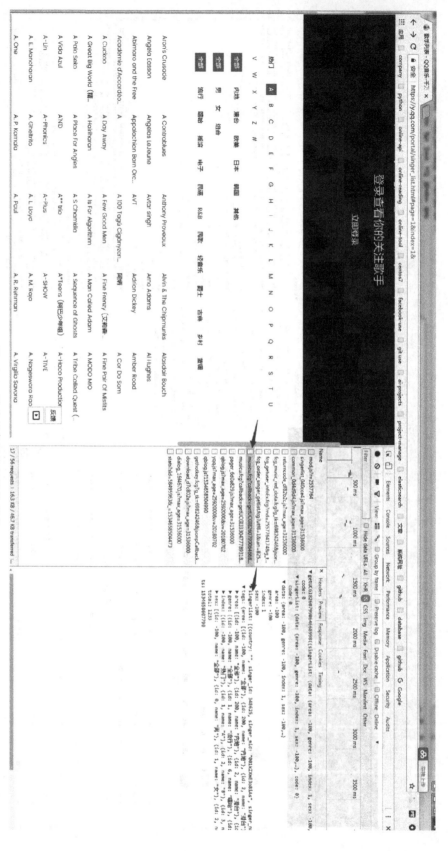

图 10-3 歌手列表的数据结构形式

根据该URL，可以进行下一步的数据查看分析，示例代码如下（url_analysis.py）：

```python
import requests
from bs4 import BeautifulSoup
import re

session = requests.session()

headers = {'Accept-Language': 'zh-CN,zh;q=0.8', 'Cache-Control': 'max-age=0',
           'User-Agent':'Mozilla/5.0 (Windows NT 6.1; WOW64) AppleWebKit/ 537.36 '
                        '(KHTML, like Gecko) Chrome/62.0.3202.94 Safari/537.36'}

QQ_MUSIC_SINGER_URL = 'https://u.y.qq.com/cgi-bin/musicu.fcg?' \
                      'callback=getUCGI1986419190708677&data=%7B%22' \
                      'comm%22%3A%7B%22ct%22%3A24%2C%22cv%22%3A10000%7D%2C%22' \
                      'singerList%22%3A%7B%22module%22%3A%22Music.SingerListServer'\
                      '%22%2C%22method%22%3A%22get_singer_list%22%2C%22' \
                      'param%22%3A%7B%22area%22%3A-100%2C%22sex%22%3A- 100%2C%22' \
                      'genre%22%3A-100%2C%22index%22%3A-100%2C%22' \
                      'sin%22%3A0%2C%22cur_page%22%3A1%7D%7D%7D'

# 取得所有字母
def get_all_letter():
    # 根据URL获取内容
    r = requests.get(QQ_MUSIC_SINGER_URL, headers=headers)
    print(QQ_MUSIC_SINGER_URL)
    all_content = r.content.decode('utf-8')
    print('all content is:\n{}'.format(all_content))
    # 获取内容，根据条件截取后，转为字典
    content_dict = eval(all_content[all_content.find("{"):len(all_content)-1])
    print('content dict is:\n{}'.format(content_dict))
    # 从字典中获取key为singerList的值
    singerList_dict = content_dict.get('singerList')
    print('singer list dict is:\n{}'.format(singerList_dict))
    # 从singerList_dict中获取key为data的值
    data_dict = singerList_dict.get('data')
    print('data dict is:\n{}'.format(data_dict))
    # 从data_dict中获取key为tags的值
    tags_dict = data_dict.get('tags')
    print('tags dict is:\n{}'.format(tags_dict))
    # 从tags_dict中获取key为index的值
    index_list = tags_dict.get('index')
    print('index list is:\n{}'.format(index_list))
    # 遍历index_list
    for item_dict in index_list:
        print('集合中的单个元素:{}'.format(item_dict))
        print('字典中的name值-------:{}'.format(item_dict.get('name')))
```

```python
# 通过解析 HTML 方式
def get_by_html():
    url = 'https://y.qq.com/portal/singer_list.html'
    r = requests.get(url, headers=headers)
    print('获取网页内容:{}'.format(r.content))
    soup = BeautifulSoup(r.content.decode('utf-8'), 'html5lib')
    print('打印 BeautifulSoup 对象:{}'.format(soup))
    # 查找所有 div 块中含 singer_tag__list js_area 标识的文本
    singer_tag_group=soup.find_all('div', re.compile('singer_tag__list js_area'))[0].find_all('a')
    print('singer tag group:{}'.format(singer_tag_group))
    # 查找所有 div 块中含 singer_tag__list js_letter 标识的文本
    alphas_group = soup.find_all('div', re.compile('singer_tag__list js_letter'))[0].find_all('a')
    print('alphas group:{}'.format(alphas_group))

# 获取字字母分类下总歌手页数
def get_all_singer():
    # 获取字母 A~Z 全部歌手
    for chr_i in range(65, 91):
        key_chr = chr(chr_i)
        # 获取每个字母分类下总歌手页数
        url = 'https://c.y.qq.com/v8/fcg-bin/v8.fcg?channel=singer&page= list&key=all_all_{}' \
              '&pagesize=100&pagenum={}&loginUin=0&hostUin=0&format= jsonp'.format(key_chr, 1)
        response = session.get(url, headers=headers)
        page_num = response.json()['data']['total_page']
        page_list = [x for x in range(page_num)]
        print('chr:{}, total page:{}'.format(chr_i, len(page_list)))

def main():
    get_all_letter()
    # get_by_html()
    # get_all_singer()

if __name__ == "__main__":
    main()
```

10.2 获取歌手的歌曲数目

有了歌手列表后,需要查看各歌手的歌曲数目及列表情况。

如在热门处单击"A",再单击"内地",选择"阿宝",得到如下 URL:https:// y.qq.com/n/yqq/singer/003oUwJ54CMqTT.html#stat=y_new.singerlist.singername

对该 URL 得到的页面使用开发者工具进行分析,如图 10-4 所示。

total 是当前歌手的全部歌曲数目。list 为歌曲列表,每页最多 30 首歌。singer_mid 是歌手的 mid,是唯一的,用于标识歌手。

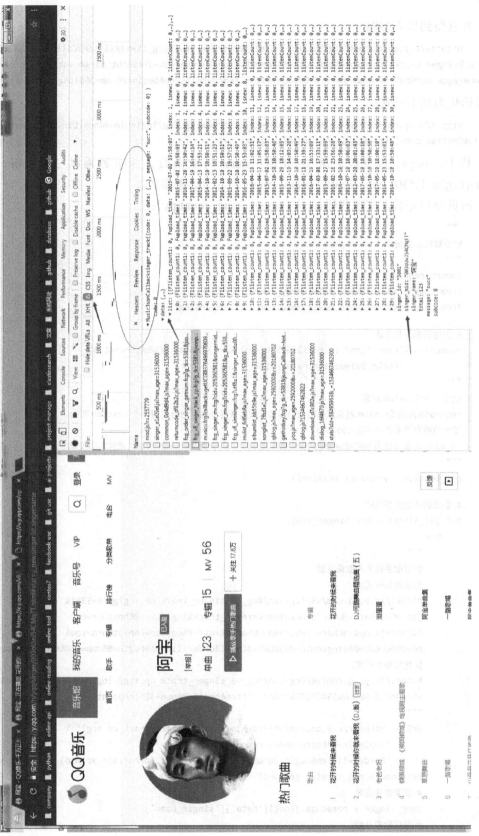

图 10-4 歌手详细页面

并且得到如下 URL：

> https://c.y.qq.com/v8/fcg-bin/fcg_v8_singer_track_cp.fcg?g_tk=5381&jsonpCallback=MusicJsonCallbacksinger_track&loginUin=0&hostUin=0&format=jsonp&inCharset=utf8&outCharset=utf-8&notice=0&platform=yqq&needNewCode=0&singermid=003oUwJ54CMqTT&order=listen&begin=0&num=30&songstatus=1

可以简化为如下形式：

> https://c.y.qq.com/v8/fcg-bin/fcg_v8_singer_track_cp.fcg?loginUin=0&hostUin=0&singermid=003oUwJ54CMqTT&order=listen&begin=0&num=30&songstatus=1

其中，begin 为页数，从 0 开始；num 为歌曲数，为页数乘以 30 所得，即 num=(begin+1)*30。singermid、begin、num 三者为动态参数。

网页分析完后，实现代码如下（singer_song_count.py）：

```python
"""
歌手歌曲总数统计
"""
import requests
from chapter10.database.sqlalchemy_conn import db_conn
from chapter10.database.music_model import SingerSong

# 创建请求头和会话
headers = {'User-Agent': 'Mozilla/5.0 (Windows NT 6.3; WOW64; rv:41.0) '
                        'Gecko/20100101 Firefox/41.0'}
"""
创建一个 Session 对象
requests 库的 Session 对象能够帮我们跨请求保持某些参数，也会在同一个 Session 实例
发出的所有请求之间保持 Cookies。
Session 对象还能为我们提供请求方法的默认数据，通过设置 Session 对象的属性来实现。
"""
req_session = requests.session()

# 获取歌手的全部歌曲
def get_singer_songs(singer_mid):
    try:
        """
        获取歌手姓名和歌曲总数
        原生地址形式：
        https://c.y.qq.com/v8/fcg-bin/fcg_v8_singer_track_cp.fcg?g_tk=5381&
        jsonpCallback=MusicJsonCallbacksinger_track&loginUin=0&hostUin=0&
        format=jsonp&inCharset=utf8&outCharset=utf-8&notice=0&platform=yqq&
        needNewCode=0&singermid=003oUwJ54CMqTT&order=listen&begin=0&num= 30&songstatus=1
        优化后地址形式：
        https://c.y.qq.com/v8/fcg-bin/fcg_v8_singer_track_cp.fcg? loginUin=0&hostUin=0&
        singermid=003oUwJ54CMqTT&order=listen&begin=0&num=30&songstatus=1
        """
        url = 'https://c.y.qq.com/v8/fcg-bin/fcg_v8_singer_track_cp.fcg?' \
              'loginUin=0&hostUin=0&singermid={}' \
              '&order=listen&begin=0&num=30&songstatus=1'.format(singer_mid)
        response = req_session.get(url)
        # 获取歌手姓名
        song_singer = response.json()['data']['singer_name']
        # 获取歌曲总数
```

```python
            song_count = response.json()['data']['total']
            print('歌手名称:{}, 歌手歌曲总数:{}'.format(song_singer, song_count))
            # 歌手歌曲总数持久化
            session_db = db_conn()
            singer_song_obj = SingerSong(singer_name=song_singer, song_count= song_count, singer_mid=singer_mid)
            session_db.add(singer_song_obj)
            session_db.commit()
            session_db.close()
        except Exception as ex:
            print('get singer info error:{}'.format(ex))

# 获取当前字母的全部歌手
def get_singer_letter(chr_key, page_list):
    for page_num in page_list:
        url = 'https://c.y.qq.com/v8/fcg-bin/v8.fcg?channel=singer&page= list&key= all_all_{}' \
              '&pagesize=100&pagenum={}&loginUin=0&hostUin=0&format=jsonp'.\
            format(chr_key, page_num + 1)
        response = req_session.get(url)
        # 循环每一个歌手
        per_singer_count = 0
        for k_item in response.json()['data']['list']:
            singer_mid = k_item['Fsinger_mid']
            get_singer_songs(singer_mid)
            per_singer_count += 1
            # 演示使用，每位歌手最多遍历5首歌
            if per_singer_count > 5:
                break
        # 演示使用，只遍历第一页
        break

# 单进程单线程方式获取全部歌手
def get_all_singer():
    # 获取字母A~Z全部歌手
    for chr_i in range(65, 91):
        key_chr = chr(chr_i)
        # 获取每个字母分类的总歌手页数
        url = 'https://c.y.qq.com/v8/fcg-bin/v8.fcg?channel=singer&page= list&key=all_all_{}' \
              '&pagesize=100&pagenum={}&loginUin=0&hostUin=0&format= jsonp'.format(key_chr, 1)
        response = req_session.get(url, headers=headers)
        page_num = response.json()['data']['total_page']
        page_list = [x for x in range(page_num)]
        # 获取当前字母下全部歌手
        get_singer_letter(key_chr, page_list)

if __name__ == '__main__':
    # 获取全部歌手
    get_all_singer()
```

执行该 PY 文件，从打印日志的控制台可以看到类似如下的输出：

```
歌手名称:Alan Walker,歌手歌曲总数:164
歌手名称:A-Lin,歌手歌曲总数:200
```

```
歌手名称:Adele,歌手歌曲总数:169
歌手名称:Avril Lavigne,歌手歌曲总数:298
歌手名称:阿杜,歌手歌曲总数:140
歌手名称:Ariana Grande,歌手歌曲总数:228
歌手名称:BIGBANG (빅뱅),歌手歌曲总数:434
歌手名称:BEYOND,歌手歌曲总数:792
歌手名称:本兮,歌手歌曲总数:166
```

通过 MySQL 客户端查看 singer_song 表，可以看到表中已插入如图 10-5 所示数据。

id	singer_name	song_count	singer_mid
1	Alan Walker	164	0020PeOh4ZaCw1
2	A-Lin	200	003ArN8Z0WpjTz
3	Adele	169	003CoxJh1zFPpx
4	Avril Lavigne	298	001Ic36m4AhaBw
5	阿杜	140	0022VtZd19rZpi
6	Ariana Grande	228	004eTCF03KvBOE
7	BIGBANG (빅뱅)	434	004AlfUb0cVkN1
8	BEYOND	792	002pUZT93gF4Cu
9	本兮	166	003LaMHm42u7qS
10	BLACKPINK	63	003DBAjk2MMfhR
11	Bruno Mars	228	001F1svH2tTpsC
12	BY2	114	000Z1Ow71FFutX
13	陈奕迅	1397	003Nz2So3XXYek
14	陈小春	274	004DFS271osAwp
15	Charlie Puth	142	000jnR7q3pCzYG

图 10-5 singer_song 表插入的数据

由插入结果可知，以上程序文件已成功从网站爬取歌手歌曲数及歌手 mid 值。

以上示例使用的是单线程处理，使用单线程做大批量数据爬取时会比较慢，耗时比较多。这里引入多线程、多进程做加速。

从 Python 3.2 开始，Python 标准库提供了 concurrent.futures 模块，其中提供了 ThreadPoolExecutor 和 ProcessPoolExecutor 两个类，对 threading 和 multiprocessing 实现了更高级的抽象，对编写线程池和进程池提供了直接的支持。

使用 ThreadPoolExecutor 和 ProcessPoolExecutor 两个类，可以将以上代码更改为如下更高效的代码形式（multi_pro_singer_song_count.py）：

```python
import math
import requests
from chapter10.database.sqlalchemy_conn import db_conn
from chapter10.database.music_model import SingerSong
from concurrent.futures import ThreadPoolExecutor, ProcessPoolExecutor

# 创建请求头和会话
headers = {'User-Agent': 'Mozilla/5.0 (Windows NT 6.3; WOW64; rv:41.0) '
           'Gecko/20100101 Firefox/41.0'}
"""
创建一个 session 对象
requests 库的 session 对象能够帮我们跨请求保持某些参数，也会在同一个 session 实例发出的所有请求之间保持 cookies
session 对象还能为我们提供请求方法的默认数据，通过设置 session 对象的属性来实现
"""
```

```python
session = requests.session()

# 获取歌手的全部歌曲
def get_singer_songs(singer_mid):
    try:
        """
        获取歌手姓名和歌曲总数
        原生地址形式:
        https://c.y.qq.com/v8/fcg-bin/fcg_v8_singer_track_cp.fcg?g_tk=5381&
        jsonpCallback=MusicJsonCallbacksinger_track&loginUin=0&hostUin=0&
        format=jsonp&inCharset=utf8&outCharset=utf-8&notice=0&platform=yqq&
        needNewCode=0&singermid=003oUwJ54CMqTT&order=listen&begin=0&num= 30&songstatus=1
        优化后地址形式:
        https://c.y.qq.com/v8/fcg-bin/fcg_v8_singer_track_cp.fcg? loginUin=0&hostUin=0&
        singermid=003oUwJ54CMqTT&order=listen&begin=0&num=30&songstatus=1
        """
        url = 'https://c.y.qq.com/v8/fcg-bin/fcg_v8_singer_track_cp.fcg? loginUin=0&' \
              'hostUin=0&singermid={}' \
              '&order=listen&begin=0&num=30&songstatus=1'.format(singer_mid)
        response = session.get(url)
        # 获取歌手姓名
        song_singer = response.json()['data']['singer_name']
        # 获取歌曲总数
        song_count = response.json()['data']['total']
        print('歌手名称:{}, 歌手歌曲总数:{}'.format(song_singer, song_count))
        # 歌手歌曲总数持久化
        session_db = db_conn()
        singer_song_obj = SingerSong(singer_name=song_singer, song_count= song_count,
                                    singer_mid=singer_mid)
        session_db.add(singer_song_obj)
        session_db.commit()
        session_db.close()
    except Exception as ex:
        print('get singer info error:{}'.format(ex))

# 获取当前字母下的全部歌手
def get_alphabet_singer(alphabet, page_list):
    for page_num in page_list:
        url = 'https://c.y.qq.com/v8/fcg-bin/v8.fcg?channel=singer&page= list&key=all_all_{}' \
              '&pagesize=100&pagenum={}&loginUin=0&hostUin=0&format=jsonp'.\
            format(alphabet, page_num + 1)
        response = session.get(url)
        # 循环每一个歌手
        per_singer_count = 0
        for k_item in response.json()['data']['list']:
            singer_mid = k_item['Fsinger_mid']
            get_singer_songs(singer_mid)
            per_singer_count += 1
            # 演示使用，每位歌手最多遍历 5 首歌
            if per_singer_count > 5:
                break
        # 演示使用，只遍历第一页
        break

# 多线程
```

```python
def multi_threading(alphabet):
    # 每个字母分类的歌手列表页数
    url = 'https://c.y.qq.com/v8/fcg-bin/v8.fcg?channel=singer&page=list&' \
          'key=all_all_{}&pagesize=100&pagenum={}&loginUin=0&hostUin= 0&format=jsonp'.\
        format(alphabet, 1)
    r = session.get(url, headers=headers)
    page_num = r.json()['data']['total_page']
    page_list = [x for x in range(page_num)]
    thread_num = 10
    # 将每个分类总页数平均分给线程数
    per_thread_page = math.ceil(page_num / thread_num)
    # 设置线程对象
    thread_obj = ThreadPoolExecutor(max_workers=thread_num)
    for thread_order in range(thread_num):
        # 计算每条线程应执行的页数
        start_num = per_thread_page * thread_order
        if per_thread_page * (thread_order + 1) <= page_num:
            end_num = per_thread_page * (thread_order + 1)
        else:
            end_num = page_num
        # 每个线程各自执行不同的歌手列表页数
        thread_obj.submit(get_alphabet_singer, alphabet, page_list[start_num: end_num])

# 多进程
def execute_process():
    with ProcessPoolExecutor(max_workers=2) as executor:
        for i in range(65, 90):
            # 创建 26 个线程，分别执行 A~Z 分类
            executor.submit(multi_threading, chr(i))

if __name__ == '__main__':
    # 执行多进程多线程
    execute_process()
```

在以上示例代码用 ThreadPoolExecutor 创建了多个线程，用 ProcessPoolExecutor 创建了多个进程。

执行以上示例代码，可以看到执行速度比单线程、单进程爬取速度提升很多。

10.3 获取每首歌曲信息

有了歌手歌曲列表后，可以进入具体的歌曲中获取每首歌曲信息。如在歌手阿宝的歌曲列表中单击"花开的时候你就来看我"这首歌，得到页面的 URL 如下：https://y.qq.com/n/yqq/song/002AOwqK2rwcrb.html。使用开发者工具分析该页面，可以得到如图 10-6 所示信息。

其中几个关键字段如下。

① songname：歌曲名称。
② albumname：歌曲所属专辑。
③ interval：歌曲时长。
④ songmid：歌曲 mid。

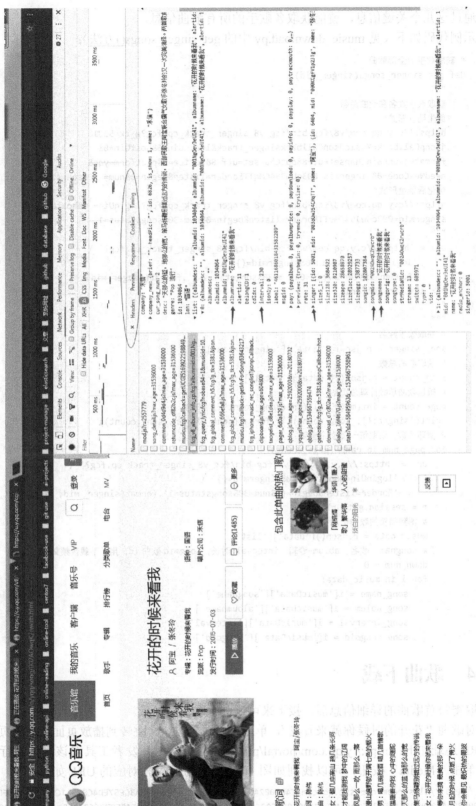

图 10-6 单首歌曲信息

通过这几个关键信息,就能获取各歌手的所有歌曲信息。

示例代码如下(见 music_download.py 中的 get_singer_songs()方法):

```python
# 获取歌手的全部歌曲
def get_singer_songs(singer_mid):
    """
    获取歌手姓名和歌曲总数
    原生地址形式:
    https://c.y.qq.com/v8/fcg-bin/fcg_v8_singer_track_cp.fcg?g_tk=5381&
    jsonpCallback=MusicJsonCallbacksinger_track&loginUin=0&hostUin=0&
    format=jsonp&inCharset=utf8&outCharset=utf-8&notice=0&platform=yqq&
    needNewCode=0&singermid=003oUwJ54CMqTT&order=listen&begin=0&num=30&songstatus=1
    优化后地址形式:
    https://c.y.qq.com/v8/fcg-bin/fcg_v8_singer_track_cp.fcg?loginUin=0&hostUin=0&
    singermid=003oUwJ54CMqTT&order=listen&begin=0&num=30&songstatus=1
    """
    url = 'https://c.y.qq.com/v8/fcg-bin/fcg_v8_singer_track_cp.fcg?' \
          'loginUin=0&hostUin=0&singermid={}' \
          '&order=listen&begin=0&num=30&songstatus=1'.format(singer_mid)
    r = session.get(url)
    """
    """
    # 获取歌手姓名
    song_singer = r.json()['data']['singer_name']
    # 获取歌曲总数
    song_count = r.json()['data']['total']
    # 根据歌曲总数计算总页数
    page_count = int(math.ceil(int(song_count) / 30))
    print('singer:{}, song count:{}'.format(song_singer, song_count))
    # 循环页数,获取每一页歌曲信息
    for page_num in range(page_count):
        url = 'https://c.y.qq.com/v8/fcg-bin/fcg_v8_singer_track_cp.fcg?' \
              'loginUin=0&hostUin=0&singermid={}' \
              '&order=listen&begin={}&num=30&songstatus=1'.format(singer_mid, page_num * 30)
        r = session.get(url)
        # 得到每页的歌曲信息
        music_data = r.json()['data']['list']
        # songname-歌名, ablum-专辑, interval-时长, songmid 歌曲 id, 用于下载音频文件
        down_num = 0
        for i in music_data:
            song_name = i['musicData']['songname']
            song_ablum = i['musicData']['albumname']
            song_interval = i['musicData']['interval']
            song_songmid = i['musicData']['songmid']
```

10.4 歌曲下载

取得每首歌曲的详细信息后,接下来可以进行歌曲的下载。

对歌曲"花开的时候你就来看我",单击"播放"按钮,跳转到播放页面,播放页面的 URL 地址如下:https://y.qq.com/portal/player.html。使用开发者工具对该网页进行分析,单击"Media"按钮,可以找到如图 10-7 所示的信息。对应的 URL 如下:

```
http://dl.stream.qqmusic.qq.com/C400002AOwqK2rwcrb.m4a?vkey=BA3CF57ED4C82311C9905BF1B5FDC1
0B74C9BE8BFA52EBD96642F03BF10AAAF9C4B2C505958260754F5A0816BEA6D15FF09548F077C6242E&guid=405034998
&uin=0&fromtag=66
```

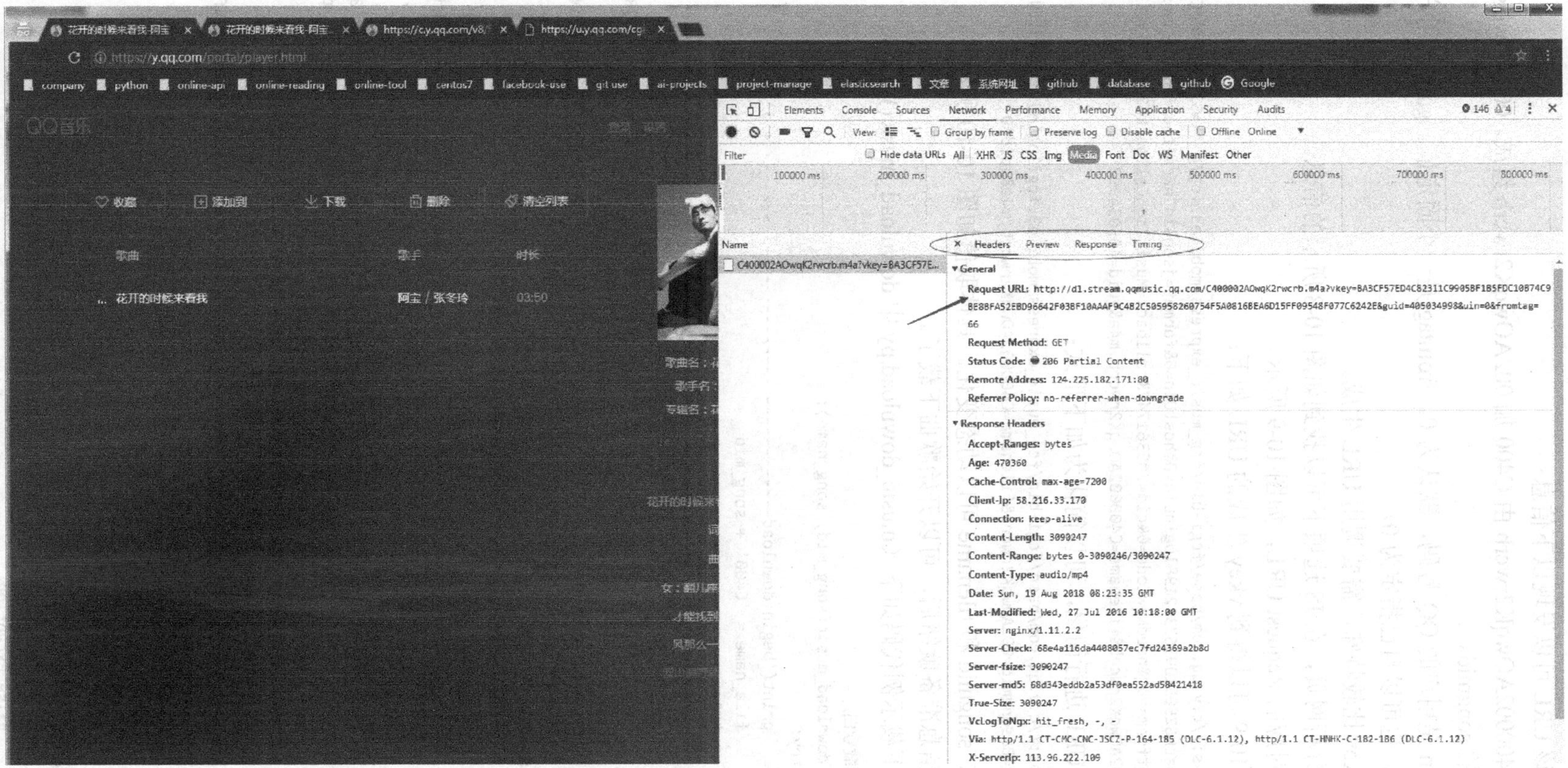

图 10-7 歌曲页面

观察该 URL 可以发现以下信息：

① C400002AOwqK2rwcrb 由 C400 加 002AOwqK2rwcrb 组成。002AOwqK2rwcrb 是歌曲对应的 songmid。

② uin 是用户的 QQ 号码，默认为 0。fromtag 是一个固定值。guid 是用户 id，由浏览器生成，也可以自己指定为 0。

③ vkey 比较特殊，需要调用 URL 生成。

继续分析网页，在 JS 选项下可以找到如图 10-8 所示的信息。

查找对应的 Request URL，如图 10-9 所示。

由图 10-9 可以得到 vkey 生成的 URL 如下：

```
https://c.y.qq.com/base/fcgi-bin/fcg_music_express_mobile3.fcg?g_tk=5381&jsonpCallback=MusicJsonCallback6250510936334219&loginUin=0&hostUin=0&format=json&inCharset=utf8&outCharset=utf-8&notice=0&platform=yqq&needNewCode=0&cid=205361747&callback=MusicJsonCallback6250510936334219&uin=0&songmid=002AOwqK2rwcrb&filename=C400002AOwqK2rwcrb.m4a&guid=405034998
```

对该 URL 进行简化，可以简化为如下形式：

```
https://c.y.qq.com/base/fcgi-bin/fcg_music_express_mobile3.fcg?loginUin=0&hostUin=0&cid=205361747&uin=0&songmid=002AOwqK2rwcrb&filename=C400002AOwqK2rwcrb.m4a&guid=0
```

其中，songmid 和 filename 可以动态变化。具体取值如图 10-10 所示。

一切信息准备就绪后，可以开始歌曲下载了。

歌曲下载示例代码如下（music_download.py 中 download_music()方法）：

```python
# 下载歌曲
def download_music(song_mid, song_name):
    try:
        print('begin download------------------------------------')
        file_name = 'C400' + song_mid
        """
        获取 vkey
        原生地址：
        https://c.y.qq.com/base/fcgi-bin/fcg_music_express_mobile3.fcg?g_tk=5381&
        jsonpCallback=MusicJsonCallback8359183970915902&loginUin= 0&hostUin=0&
        format=json&inCharset=utf8&outCharset=utf-8&notice=0&platform= yqq&needNewCode=0&
        cid=205361747&callback=MusicJsonCallback8359183970915902&uin=0&songmid=002AOwqK2rwcrb&
        filename=C400002AOwqK2rwcrb.m4a&guid=3192684595
        优化后地址：
        https://c.y.qq.com/base/fcgi-bin/fcg_music_express_mobile3.fcg?loginUin=0&hostUin=0&
        cid=205361747&uin=0&songmid=002AOwqK2rwcrb&filename= C400002AOwqK2rwcrb.m4a&guid=0
        """
        vkey_url='https://c.y.qq.com/base/fcgi-bin/fcg_music_express_mobile3.fcg?loginUin=0&hostUin=0'\
                 '&cid=205361747&uin=0&songmid={}&filename={}.m4a&guid=0'.format(song_mid,file_name)
        vkey_response = session.get(vkey_url, headers=headers)
        vkey = vkey_response.json()['data']['items'][0]['vkey']
        # 下载歌曲
        url = 'http://dl.stream.qqmusic.qq.com/{}.m4a?vkey={}&guid=0&uin= 0&fromtag=66'.format(file_name, vkey)
        response = session.get(url, headers=headers)
        music_download_path = download_path + '/{}.m4a'.format(song_name)
        with open(music_download_path, 'wb') as f_write:
            f_write.write(response.content)
    except Exception as ex:
        print('download error:{}'.format(ex))
        raise Exception
```

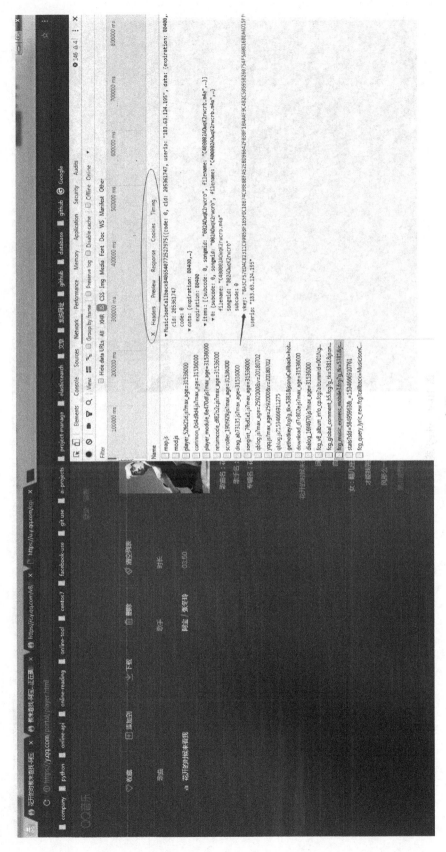

图 10-8 vkey 查找

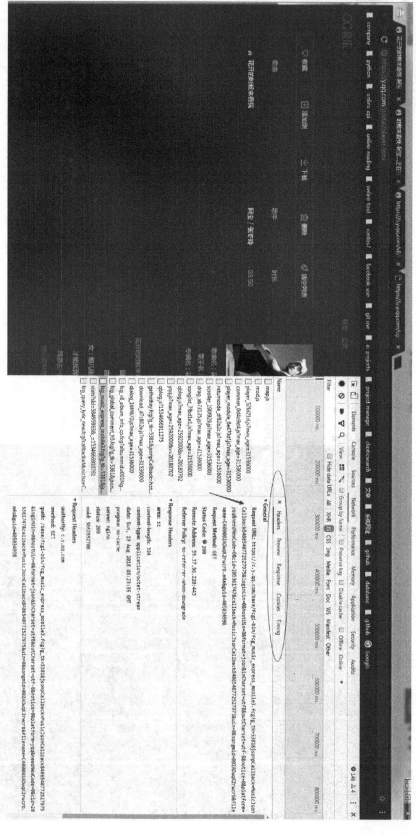

图 10-9 vkey 获取 URL

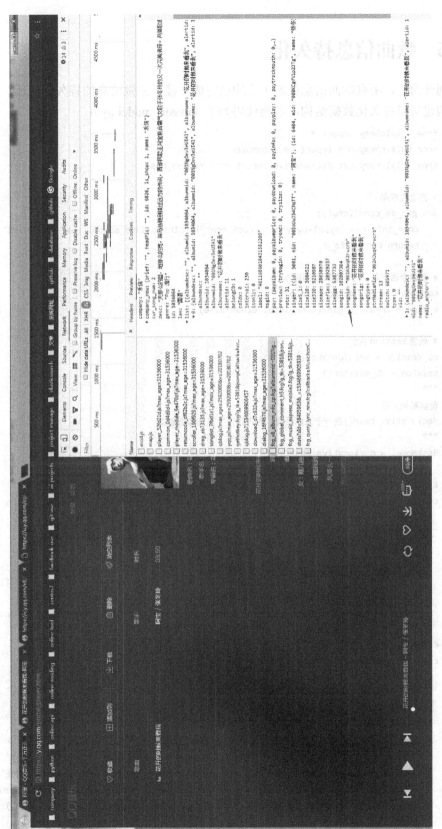

图 10-10 具体字段信息获取

10.5 歌曲信息持久化

到此为止，所有歌曲信息处理工作均已完成，接下来做信息的持久化处理。

构建信息持久化数据结构，示例代码如下（music_model.py）：

```python
from sqlalchemy import *
from sqlalchemy.orm import sessionmaker
from sqlalchemy.ext.declarative import declarative_base

# 连接数据库
def get_db_conn_info():
    conn_info_r = "mysql+pymysql://root:root@localhost/data_school?charset = UTF8"
    return conn_info_r

# 创建会话对象，用于数据表的操作
conn_info = get_db_conn_info()
engine = create_engine(conn_info, echo=False)

# 创建 Session 实例
db_session = sessionmaker(bind=engine)
session = db_session()
"""
创建基类实例
declarative_base()是一个工厂函数，为声明性类定义构造基类
"""
BaseModel = declarative_base()

# 映射数据表
class SingerSong(BaseModel):
    # 表名
    __tablename__ ='singer_song'
    # 字段，属性
    id = Column(Integer, primary_key=True)
    singer_name = Column(String(100), default=None, nullable=True, comment= '歌手名称')
    song_count = Column(Integer, default=0, nullable=False, comment='歌曲数')
    singer_mid = Column(String(100), default=None, nullable=True, comment= '歌手 mid')

# 映射数据表
class Song(BaseModel):
    # 表名
    __tablename__ ='song'
    # 字段，属性
    id = Column(Integer, primary_key=True)
    song_name = Column(String(50), default=None, nullable=True, comment=  '歌名')
    song_ablum = Column(String(50), default=None, nullable=True, comment= '专辑')
    song_interval = Column(String(50), default=None, nullable=True, comment= '时长')
    song_songmid = Column(String(50), default=None, nullable=True, comment= '歌曲 mid')
    song_singer = Column(String(50), default=None, nullable=True, comment= '歌手')
```

```python
# 创建数据表
BaseModel.metadata.create_all(engine)
```

数据库连接代码封装，示例代码如下（sqlalchemy_conn.py）：

```python
from sqlalchemy import create_engine
from sqlalchemy.orm import sessionmaker

# 数据库连接
def db_conn():
    conn_info = "mysql+pymysql://root:root@localhost/data_school?charset= utf8"
    engine = create_engine(conn_info, echo=False)
    db_session = sessionmaker(bind=engine)
    session = db_session()
    return session

# 数据库查询
def query_mysql(sql_str):
    session = db_conn()
    return session.execute(sql_str)

# 数据库更新
def update_mysql(update_sql):
    session = db_conn()
    session.execute(update_sql)
    session.commit()
    session.close()

if __name__ == "__main__":
    sql = 'select * from song'
    result = query_mysql(sql)
    for item in result:
        print(item)
```

配置文件示例代码（config.py）：

```python
import os

# 根路径
ROOT_PATH = os.getcwd()
```

歌曲下载示例代码（music_download.py）：

```python
import math
import os
import requests

from chapter10.database.sqlalchemy_conn import db_conn
from concurrent.futures import ThreadPoolExecutor, ProcessPoolExecutor
from chapter10.database.music_model import Song
from chapter10.config import ROOT_PATH

download_path = os.path.join(ROOT_PATH, 'music_file')
```

```python
# 创建请求头和会话
headers = {'User-Agent': 'Mozilla/5.0 (Windows NT 6.3; WOW64; rv:41.0) '
                        'Gecko/20100101 Firefox/41.0'}
"""
创建一个 Session 对象
requests 库的 Session 对象能够帮我们跨请求保持某些参数，也会在同一个 Session 实例
发出的所有请求之间保持 Cookies。
Session 对象还能为我们提供请求方法的默认数据，通过设置 Session 对象的属性来实现。
"""
session = requests.session()

# 下载歌曲
def download_music(song_mid, song_name):
    try:
        print('begin download------------------------------------')
        file_name = 'C400' + song_mid
        """
        获取 vkey
        原生地址：
        https://c.y.qq.com/base/fcgi-bin/fcg_music_express_mobile3.fcg?g_tk=5381&
        jsonpCallback=MusicJsonCallback8359183970915902&loginUin=0&hostUin=0&
        format=json&inCharset=utf8&outCharset=utf-8&notice=0&platform=yqq&needNewCode=0&
        cid=205361747&callback=MusicJsonCallback8359183970915902&uin=0&songmid=002AOwqK2rwcrb&
        filename=C400002AOwqK2rwcrb.m4a&guid=3192684595
        优化后地址：
        https://c.y.qq.com/base/fcgi-bin/fcg_music_express_mobile3.fcg?loginUin=0&hostUin=0&
        cid=205361747&uin=0&songmid=002AOwqK2rwcrb&filename= C400002AOwqK2rwcrb.m4a&guid=0
        """
        vkey_url = 'https://c.y.qq.com/base/fcgi-bin/fcg_music_express_ mobile3.fcg?' \
                   'loginUin=0&hostUin=0&cid=205361747&uin=0&songmid={}' \
                   '&filename={}.m4a&guid=0'.format(song_mid, file_name)
        vkey_response = session.get(vkey_url, headers=headers)
        vkey = vkey_response.json()['data']['items'][0]['vkey']
        # 下载歌曲
        url = 'http://dl.stream.qqmusic.qq.com/{}.m4a?vkey={}&guid=0&uin= 0&fromtag=66'.\
            format(file_name, vkey)
        response = session.get(url, headers=headers)
        music_download_path = download_path + '/{}.m4a'.format(song_name)
        with open(music_download_path, 'wb') as f_write:
            f_write.write(response.content)
    except Exception as ex:
        print('download error:{}'.format(ex))
        raise Exception

# 获取歌手的全部歌曲
def get_singer_songs(singer_mid):
    """
    获取歌手姓名和歌曲总数
    原生地址形式：
    https://c.y.qq.com/v8/fcg-bin/fcg_v8_singer_track_cp.fcg?g_tk=5381&
    jsonpCallback=MusicJsonCallbacksinger_track&loginUin=0&hostUin=0&
    format=jsonp&inCharset=utf8&outCharset=utf-8&notice=0&platform=yqq&
```

```python
needNewCode=0&singermid=003oUwJ54CMqTT&order=listen&begin=0&num=30& songstatus=1
优化后地址形式:
https://c.y.qq.com/v8/fcg-bin/fcg_v8_singer_track_cp.fcg?loginUin= 0&hostUin=0&
singermid=003oUwJ54CMqTT&order=listen&begin=0&num=30&songstatus=1
"""
url = 'https://c.y.qq.com/v8/fcg-bin/fcg_v8_singer_track_cp.fcg?' \
      'loginUin=0&hostUin=0&singermid={}' \
      '&order=listen&begin=0&num=30&songstatus=1'.format(singer_mid)
r = session.get(url)
"""
"""
# 获取歌手姓名
song_singer = r.json()['data']['singer_name']
# 获取歌曲总数
song_count = r.json()['data']['total']
# 根据歌曲总数计算总页数
page_count = int(math.ceil(int(song_count) / 30))
print('singer:{}, song count:{}'.format(song_singer, song_count))
# 循环页数，获取每一页歌曲信息
for page_num in range(page_count):
    url = 'https://c.y.qq.com/v8/fcg-bin/fcg_v8_singer_track_cp.fcg?' \
          'loginUin=0&hostUin=0&singermid={}' \
          '&order=listen&begin={}&num=30&songstatus=1'.format(singer_mid, page_num * 30)
    r = session.get(url)
    # 得到每页的歌曲信息
    music_data = r.json()['data']['list']
    # songname-歌名，ablum-专辑，interval-时长，songmid 歌曲 id，用于下载音频文件
    down_num = 0
    for i in music_data:
        song_name = i['musicData']['songname']
        song_ablum = i['musicData']['albumname']
        song_interval = i['musicData']['interval']
        song_songmid = i['musicData']['songmid']
        # 下载歌曲
        download_music(song_songmid, song_name)
        # 入库处理
        song_obj = Song(song_name=song_name, song_ablum=song_ablum,
                        song_interval=song_interval,song_songmid= song_songmid,
                        song_singer=song_singer)
        session_db = db_conn()
        session_db.add(song_obj)
        session_db.commit()
        session_db.close()
        down_num += 1
        # 为演示使用，此处示例下载 5 首歌曲
        if down_num > 5:
            break

# 获取当前字母的全部歌手
def get_alphabet_singer(alphabet, page_list):
    for page_num in page_list:
        url = 'https://c.y.qq.com/v8/fcg-bin/v8.fcg?channel=singer&page= list&key=all_all_{}' \
```

```python
                    '&pagesize=100&pagenum={}&loginUin=0&hostUin=0&format=jsonp'.\
                        format(alphabet, page_num + 1)
            response = session.get(url)
            # 循环每一个歌手
            for k in response.json()['data']['list']:
                singer_mid = k['Fsinger_mid']
                get_singer_songs(singer_mid)

# 多线程
def multi_threading(alphabet):
    # 每个字母分类的歌手列表页数
    url = 'https://c.y.qq.com/v8/fcg-bin/v8.fcg?channel=singer&page=list&' \
          'key=all_all_{}&pagesize=100&pagenum={}&loginUin=0&hostUin= 0&format= jsonp'.\
        format(alphabet, 1)
    response = session.get(url, headers=headers)
    page_num = response.json()['data']['total_page']
    page_list = [x for x in range(page_num)]
    thread_num = 5
    # 将每个分类总页数平均分给线程数
    per_thread_page = math.ceil(page_num / thread_num)
    # 设置线程对象
    thread_obj = ThreadPoolExecutor(max_workers=thread_num)
    for thread_order in range(thread_num):
        # 计算每条线程应执行的页数
        start_num = per_thread_page * thread_order
        if per_thread_page * (thread_order + 1) <= page_num:
            end_num = per_thread_page * (thread_order + 1)
        else:
            end_num = page_num
        # 每个线程各自执行不同的歌手列表页数
        thread_obj.submit(get_alphabet_singer, alphabet, page_list[start_num: end_num])

# 多进程
def execute_process():
    with ProcessPoolExecutor(max_workers=2) as executor:
        for i in range(65, 91):
            # 创建26个线程，分别执行A~Z分类
            executor.submit(multi_threading, chr(i))

if __name__ == '__main__':
    # 执行多进程多线程
    execute_process()
```

执行以上代码，在日志打印控制台可以看到类似如下的输出信息：

```
singer:王力宏, song count:440
begin download-------------------------------------
begin download-------------------------------------
begin download-------------------------------------
singer:Stuttgart Radio Symphony Orchestra, song count:1715
begin download-------------------------------------
begin download-------------------------------------
```

```
singer:US Air Force Heartland of America Band, song count:0
singer:Unnerbuss13, song count:2
begin download------------------------------------
```

通过 MySQL 客户端查看 song 表,可以看到表中已插入如图 10-11 所示的数据。

图 10-11　song 表插入的数据

在文件夹 chapter10/music_file 下可以看到下载的音频文件,如图 10-12 所示。

图 10-12　下载的音频文件

该示例代码中直接使用了多线程、多进程的处理方式。

本章的示例代码使用的是将数据保存到关系型数据库(MySQL)的方式。这里也可以更改为保存到非关系型数据库(MongoDB)的方式。

要修改为保存到 MongoDB,更改比较大的一个地方是获取数据库和集合,更改的示例代码如下(mongo_conn.py):

```
import pymongo

def get_mongo_conn():
    """
    取得数据库连接
    :return: 指定数据库
    """
    mongo_client = pymongo.MongoClient("mongodb://localhost:27017/")
    # 数据库连接,若存在 data_school,则直接连接,否则创建
```

```python
        mongo_conn = mongo_client["data_school"]
        return mongo_conn

    def col_singer_song():
        """
        取得指定集合
        :return: 指定集合
        """
        mongo_conn = get_mongo_conn()
        # 集合获取,若存在 singer_song 集合,则直接返回,否则创建
        mongo_col = mongo_conn["singer_song"]
        return mongo_col

    def col_song():
        """
        取得指定集合
        :return: 指定集合
        """
        mongo_conn = get_mongo_conn()
        # 集合获取,若存在 song 集合,则直接返回,否则创建
        mongo_col = mongo_conn["song"]
        return mongo_col
```

相对于前面章节的 MongoDB 的连接方式,该示例代码中需要获取两个集合,所以需要两个集合获取的方法。要获取多个集合的情况,也可以使用该方法。

获取歌手歌曲数代码更改比较少,相对于关系型数据库保存数据的方式,只是对数据保存方式做了修改,此处就不详细展示代码了,具体可以参考示例代码(mongo_singer_song_count.py 和 mongo_multi_singer_song_count.py)。

获取每首歌曲信息及歌曲下载的代码更改也比较少,只是对数据保存方式做了修改,此处就不详细展示代码了,具体可以参考示例代码(mongo_music_download.py)。

10.6 可视化展示

经过前面的一番操作后,可以获得 QQ 音乐网站几乎所有的数据,有了数据后,就可以构建可视化数据报表。结合第 9 章的知识,下面通过可视化方式展示歌手歌曲数。

10.2 节的操作已经获取了歌手及歌手歌曲数,并把结果存放到 singer_song 表,要做歌手歌曲数的可视化,操作 singer_song 表即可。我们从 singer_song 表查询出 singer_name 和 song_count 两个字段的值进行可视化展示。

借助 sqlalchemy 从 singer_song 表获取结果的操作方式如下:

```python
    def query_from_mysql():
        """
        从表中查询结果
        :return: tuple 集列表
        """
        # 从 singer_song 表查询 singer_name, song_count 字段,为方便展示,查询 10 条
        query_result = db_conn().query(SingerSong.singer_name, SingerSong.song_count).limit(10).all()
        return query_result
```

使用水平图表、饼图等展示结果的操作与第9章类似，此处不再具体阐述。

完整的示例代码如下（info_visible.py）：

```python
# 信息可视化
import os
from chapter10.database.sqlalchemy_conn import db_conn
from pyecharts.charts import Bar, Pie, WordCloud
from pyecharts import options as opts
from chapter10.config import ROOT_PATH
from chapter10.database.music_model import SingerSong

file_path_pre = os.path.join(ROOT_PATH, 'static/')

# 关键词统计字典
data_pairs = []

def query_from_mysql():
    """
    从表中查询结果
    :return: tuple 集列表
    """
    # 从 singer_song 表查询 singer_name, song_count 字段，为方便展示，查询 10 条
    query_result = db_conn().query(SingerSong.singer_name, SingerSong.song_count).limit(10).all()
    return query_result

# 结果可视化
def result_visible():
    word_tuple_list = query_from_mysql()
    for word_tuple in word_tuple_list:
        if word_tuple is None or len(word_tuple) <= 0:
            continue

        data_pairs.append(word_tuple)

    draw_bar_horizontal()
    draw_pie()
    draw_word_cloud()

# 水平图表
def draw_bar_horizontal():
    word_keys = [k for k, v in data_pairs]
    word_values = [v for k, v in data_pairs]
    bar = Bar()
    bar.add_xaxis(word_keys)
    bar.add_yaxis('歌手名', word_values)
    bar.set_global_opts(title_opts=opts.TitleOpts(title='水平图表', subtitle='歌手歌曲数'))
    # HTML 文件存放路径
    bar.render(path=file_path_pre + "bar_horizontal.html")

# 饼图
def draw_pie():
```

```
        pie = Pie()
    pie.add("歌手名", data_pairs)
    pie.set_global_opts(title_opts=opts.TitleOpts(title='饼图'))
        # HTML 文件存放路径
    pie.render(path=file_path_pre + "pie.html")

    # 词云图
    def draw_word_cloud():
    word_cloud = WordCloud()
    word_cloud.add("", data_pairs, word_size_range=[20, 100])
    word_cloud.set_global_opts(title_opts=opts.TitleOpts(title='词云图'))
        # HTML 文件存放路径
    word_cloud.render(path=file_path_pre + "word_cloud.html")

    if __name__ == "__main__":
    result_visible()
```

执行以上示例代码，在根目录的 static 目录中会生成几个 HTML 文件。用浏览器打开这些 HTML 文件，可以看到对应的可视化图表，如图 10-13 所示。

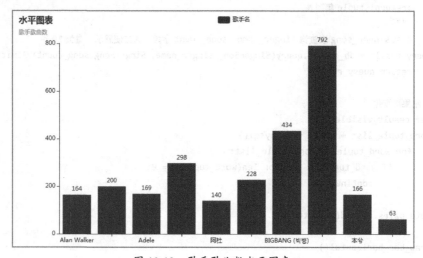

图 10-13 歌手歌曲数水平图表

本实例的操作方式与第 9 章类似，饼图及词云图的展示效果就不展示，读者可以亲自操作查看相关效果，也可以根据自己的喜好设计更多的可视化展示效果。

10.7 小结

本章通过爬取 QQ 音乐网站的数据，对前面所学内容进一步巩固，同时通过该项目实现并了解网站数据的爬取。

本章的操作示例中只展示了爬取歌手歌曲数、歌手部分信息等基础信息及歌曲文件下载的实现方式。

对于本章展示的项目，读者可以扩展为一个可以获取更多信息的项目，可以做更细一步信息的爬取，如歌曲播放时间长度、歌曲歌词、歌曲图片等信息，都可以修改后获取。

附录 A MySQL 的四个默认库

终端登录 MySQL 数据库,输入 show databases 指令显示全部数据库(或者直接用客户端工具展示),结果如下:

```
mysql> show databases;
+--------------------+
| Database           |
+--------------------+
| information_schema |
| mysql              |
| performance_schema |
| sys                |
+--------------------+
4 rows in set (0.00 sec)
```

MySQL 5.7 及以上版本的自带库为 information_schema、mysql、performance_schema、sys。MySQL 5.6 及以下版本的自带库为 information_schema、mysql、performance_schema、test。

四个库的作用分别介绍如下。

1. information_schema

information_schema 提供了访问数据库元数据的方式(元数据是关于数据的数据,如数据库名或表名、列的数据类型或访问权限等。有时用于表述该信息的其他术语包括"数据词典"和"系统目录")。

换句换说,information_schema 是一个信息数据库,保存着关于 MySQL 服务器所维护的所有其他数据库的信息(如数据库名、数据库的表、表栏的数据类型与访问权限等)。information_schema 中有几张只读表,它们实际上是视图,而不是基本表。

在终端操作如下,得到相关结果:

```
mysql> use information_schema
Database changed
mysql> show tables;
+---------------------------------------+
| Tables_in_information_schema          |
+---------------------------------------+
| CHARACTER_SETS                        |
| COLLATION_CHARACTER_SET_APPLICABILITY |
| COLLATIONS                            |
| COLUMN_PRIVILEGES                     |
| COLUMN_STATISTICS                     |
| COLUMNS                               |
| ENGINES                               |
| EVENTS                                |
| FILES                                 |
```

```
| INNODB_BUFFER_PAGE             |
| INNODB_BUFFER_PAGE_LRU         |
| INNODB_BUFFER_POOL_STATS       |
| INNODB_CACHED_INDEXES          |
| INNODB_CMP                     |
| INNODB_CMP_PER_INDEX           |
| INNODB_CMP_PER_INDEX_RESET     |
| INNODB_CMP_RESET               |
| INNODB_CMPMEM                  |
| INNODB_CMPMEM_RESET            |
| INNODB_COLUMNS                 |
| INNODB_DATAFILES               |
| INNODB_FIELDS                  |
| INNODB_FOREIGN                 |
| INNODB_FOREIGN_COLS            |
| INNODB_FT_BEING_DELETED        |
| INNODB_FT_CONFIG               |
| INNODB_FT_DEFAULT_STOPWORD     |
| INNODB_FT_DELETED              |
| INNODB_FT_INDEX_CACHE          |
| INNODB_FT_INDEX_TABLE          |
| INNODB_INDEXES                 |
| INNODB_METRICS                 |
| INNODB_TABLES                  |
| INNODB_TABLESPACES             |
| INNODB_TABLESPACES_BRIEF       |
| INNODB_TABLESTATS              |
| INNODB_TEMP_TABLE_INFO         |
| INNODB_TRX                     |
| INNODB_VIRTUAL                 |
| KEY_COLUMN_USAGE               |
| KEYWORDS                       |
| OPTIMIZER_TRACE                |
| PARAMETERS                     |
| PARTITIONS                     |
| PLUGINS                        |
| PROCESSLIST                    |
| PROFILING                      |
| REFERENTIAL_CONSTRAINTS        |
| RESOURCE_GROUPS                |
| ROUTINES                       |
| SCHEMA_PRIVILEGES              |
| SCHEMATA                       |
| ST_GEOMETRY_COLUMNS            |
| ST_SPATIAL_REFERENCE_SYSTEMS   |
| STATISTICS                     |
| TABLE_CONSTRAINTS              |
| TABLE_PRIVILEGES               |
| TABLES                         |
| TABLESPACES                    |
| TRIGGERS                       |
| USER_PRIVILEGES                |
```

```
        | VIEWS                                |
        +--------------------------------------+
        62 rows in set (0.01 sec)
```

对上述展示结果中部分表的说明如下。

① CHARACTER_SETS（字符集）表：提供了 MySQL 实例可用字符集的信息。SHOW CHARACTER SET 结果集取自此表。

② COLLATION_CHARACTER_SET_APPLICABILITY 表：指明了可用于校对的字符集。这些列等效于 SHOW COLLATION 的前两个显示字段。

③ COLLATIONS 表：提供了关于各字符集的对照信息。

④ COLUMN_PRIVILEGES（列权限）表：给出了关于列权限的信息。该信息源自 mysql.columns_priv 授权表，是非标准表。

⑤ COLUMNS 表：提供了表中的列信息，详细描述了某张表的所有列及每列的信息。SHOW COLUMNS FROM SCHEMANAME.TABLENAME 的结果取自此表。

⑥ KEY_COLUMN_USAGE 表：描述了具有约束的键列。

⑦ ROUTINES 表：提供了关于存储子程序（存储程序和函数）的信息，此时 ROUTINES 表不包含自定义函数（UDF）。名为 "mysql.proc name" 的列指明了对应 INFORMATION_SCHEMA.ROUTINES 表的 mysql.proc 表列。

⑧ SCHEMA_PRIVILEGES（方案权限）表：给出了关于方案（数据库）权限的信息。该信息来自 mysql.db 授权表，非标准表。

⑨ SCHEMATA 表：提供了当前 MySQL 实例中所有数据库的信息。SHOW DATABASES 的结果取自此表。

⑩ STATISTICS 表：提供了关于表索引的信息。SHOW INDEX FROM SCHEMANAME.TABLENAME 的结果取自此表。

⑪ TABLE_CONSTRAINTS 表：描述了存在约束的表以及表的约束类型。

⑫ TABLE_PRIVILEGES（表权限）表：给出了关于表权限的信息。该信息源自 mysql.tables_priv 授权表，非标准表。

⑬ TABLES 表：提供了关于数据库中的表的信息（包括视图），详细描述了某个表属于哪个 schema、表类型、表引擎、创建时间等信息。SHOW TABLES FROM SCHEMANAME 的结果取自此表。

⑭ TRIGGERS 表：提供了关于触发程序的信息，必须有 SUPER 权限才能查看。

⑮ USER_PRIVILEGES（用户权限）表：给出了关于全程权限的信息。该信息源自 mysql.user 授权表，非标准表。

⑯ VIEWS 表：给出了关于数据库中的视图的信息，需要有 SHOW VIEWS 权限，否则无法查看视图信息。

2. mysql

在终端操作如下，得到相关结果：

```
mysql> USE mysql
Database changed
mysql> SHOW tables;
        +---------------------------+
        | Tables_in_mysql           |
        +---------------------------+
```

```
| columns_priv              |
| component                 |
| db                        |
| default_roles             |
| engine_cost               |
| func                      |
| general_log               |
| global_grants             |
| gtid_executed             |
| help_category             |
| help_keyword              |
| help_relation             |
| help_topic                |
| innodb_index_stats        |
| innodb_table_stats        |
| password_history          |
| plugin                    |
| procs_priv                |
| proxies_priv              |
| role_edges                |
| server_cost               |
| servers                   |
| slave_master_info         |
| slave_relay_log_info      |
| slave_worker_info         |
| slow_log                  |
| tables_priv               |
| time_zone                 |
| time_zone_leap_second     |
| time_zone_name            |
| time_zone_transition      |
| time_zone_transition_type |
| user                      |
+---------------------------+
33 rows in set (0.00 sec)
```

mysql 库是核心数据库,类似 SQL Server 中的 master 表,主要负责存储数据库的用户、权限设置、关键字等需要的控制和管理信息(常用的是在 mysql.user 表中修改 root 用户的密码)。

3. performance_schema

performance_schema 库主要用于收集数据库服务器性能参数,并且库里表的存储引擎均为 PERFORMANCE_SCHEMA,用户不能创建存储引擎为 PERFORMANCE_SCHEMA 的表。MySQL 5.7 默认是开启的。

4. sys

sys 库所有的数据源来自 performance_schema,目标是把 performance_schema 的复杂度降低,让 DBA 能更好地阅读该库的内容,更快地了解数据库的运行情况。

5. test(5.6 及以下版本)

test 是安装时创建的一个测试数据库,是一个空数据库,没有任何表,可以删除。

附录 B PyMySQL 连接对象全量参数解释

host：数据库服务器所在的主机。

user：以登录身份登录的用户名。

password：要使用的密码。

database：要使用的数据库，None 不使用特定的数据库。

port：要使用的 MySQL 端口，通常默认（默认值：3306）。

bind_address：当客户端具有多个网络接口时，请指定从中连接到主机的接口。参数可以是主机名或 IP 地址。

unix_socket：选择使用 UNIX 套接字而不是 TCP/IP。

read_timeout：以秒为单位读取连接的超时（默认值，无：无超时）。

write_timeout：以秒为单位写入连接的超时（默认值，无：无超时）。

charset：要使用的 Charset。

sql_mode：要使用的默认 SQL_MODE。

read_default_file：指定 my.cnf 文件，以从 client 部分读取这些参数。

conv：使用转换字典而不是默认字典，用于提供类型的自定义编组和解组，见转换器。

use_unicode：是否默认为 Unicode 字符串。对于 Py3k，此选项默认为 true。

client_flag：要发送给 MySQL 的自定义标志，在 constants.CLIENT 中查找潜在值。

cursorclass：要使用的自定义游标类。

init_command：建立连接时要运行的初始 SQL 语句。

connect_timeout：连接时抛出异常之前的超时（默认值：10，最小值：1，最大值：31536000）。

ssl：类似 mysql_ssl_set()参数的 dict。

read_default_group：要在配置文件中读取的组。

compress：不支持。

named_pipe：不支持。

autocommit：自动提交模式。若为无，则表示使用服务器默认值（默认值：false）。

local_infile：允许使用 LOAD DATA LOCAL 命令的布尔值（默认值：false）。

max_allowed_packet：发送到服务器的最大数据包大小（以字节为单位）（默认值：16MB），仅用于限制小于默认值（16KB）的"LOAD LOCAL INFILE"数据包的大小。

defer_connect：不要明确连接 contruction——等待连接调用（默认值：false）。

auth_plugin_map：插件名称的一个字典，用于处理该插件的类。该类将 Connection 对象作为构造函数的参数，需要一个认证方法，将认证包作为参数。对话框插件可以使用提示（echo，prompt）方法（如果没有 authenticate 方法）从用户返回字符串（实验）。

server_public_key：SHA256 authenticnticaiton 插件公钥值（默认：无）。

db：数据库的别名（与 MySQL 兼容）。

passwd：密码的别名（与 MySQL 兼容）。

binary_prefix：在字节和 bytearray 上添加_binary 前缀（默认值：false）。

反侵权盗版声明

电子工业出版社依法对本作品享有专有出版权。任何未经权利人书面许可，复制、销售或通过信息网络传播本作品的行为；歪曲、篡改、剽窃本作品的行为，均违反《中华人民共和国著作权法》，其行为人应承担相应的民事责任和行政责任，构成犯罪的，将被依法追究刑事责任。

为了维护市场秩序，保护权利人的合法权益，本社将依法查处和打击侵权盗版的单位和个人。欢迎社会各界人士积极举报侵权盗版行为，本社将奖励举报有功人员，并保证举报人的信息不被泄露。

举报电话：（010）88254396；（010）88258888
传　　真：（010）88254397
E-mail：dbqq@phei.com.cn
通信地址：北京市海淀区万寿路 173 信箱
　　　　　电子工业出版社总编办公室
邮　　编：100036